Understanding OPNFV

Accelerate NFV Transformation using OPNFV

Amar Kapadia
Nicholas Chase

Understanding OPNFV

Copyright © 2017 Mirantis, Inc.

ISBN: 978-1545442708

All rights reserved. No part of the book may be reproduced, stored in a retrieval system, or transmitted in any form or by any means, without the prior written permission of the publisher, except in the case of brief quotations embedded in critical articles or reviews.

Every effort has been made in the preparation of this book to ensure the accuracy of the information presented. However, the information contained in this book is sold without warranty, either express or implied. Neither the authors, nor Mirantis, Inc. and its dealers and distributors will be held liable for any damages caused or alleged to be caused directly or indirectly by this book.

Mirantis, Inc. has endeavored to provide trademark information about all the companies and products mentioned in this book by the appropriate use of appropriate marks. However, Mirantis, Inc. cannot guarantee the accuracy of this information.

First publication: April 2017
Second publication: May 2017

Published by Mirantis Inc.
525 Almanor Avenue
Sunnyvale, CA 94085, U.S.A.
www.mirantis.com

Credits

Authors

Amar Kapadia
Nicholas Chase

Reviewers

Bryan Sullivan
Brandon Wick
Christopher Price
Fatih Degirmenci
Frank Brockners
Gregory Elkinbard
Heather Kirksey
Kathy Cacciatore
Marc Cohn
Morgan Richomme
Raymond Paik
Serg Melikyan
Steve Vandris
Tapio Tallgren
Yongsheng Gong

Interviewees

Christopher Price
Dusty Robbins
Madhu Kashyap
Prakash Ramachandran
Susan James

Cover Work

David Stoltenberg

Graphics

Amar Kapadia

Editor & Proofreader

Nicholas Chase

About the Authors

Amar Kapadia is an NFVI specialist consulting for Mirantis, the pure play open cloud company, and a NFV Strategy Orchestrator (pun intended) at Aarna Networks, Inc, his current startup gig. He was previously senior director of product marketing at Mirantis. Prior to Mirantis, he was the senior director for EVault's Long-Term Storage Service, a public cloud storage offering based on OpenStack Swift. He has over 20 years of experience in storage, server and networking technologies at Emulex, Philips and HP. Amar's current passion is in NFV, unikernels, containers, serverless scheduling, OPNFV, ONAP and CORD. He holds a master's degree in electrical engineering from the University of California, Berkeley. His blogs can be found at http://aarnanetworks.com/blog, or you can reach him on Twitter at @akapadia_usa.

> I would like to thank Boris Renski from Mirantis and Adarsh Sogal from Intel for funding this book. I want to especially thank Brandon Wick from the Linux Foundation; without his encouragement and community outreach this book would not have been possible. Finally, I would like to thank my family for tolerating my late-night and weekend book writing sessions.

Nicholas Chase is Head of Technical and Marketing Content for Mirantis, the pure play open cloud company. Prior to Mirantis, he was an IBM DeveloperWorks Master Author, the Chief Technical Officer of an interactive development company, an Oracle Instructor, a software developer, the founder of the social bookmarking site NoTooMi.com, author or co-author of more than a dozen technical books, a sysop on the original Cleveland Freenet, and the driving force behind one of the very first applications to use a web browser to update a database. He lives on a farm with his wonderful wife Sarah Jane, 27 chickens, 12 goats, 7 ducks, 2 dogs and a furbaby, 3 fish, 2 parakeets, and Hazel the Hedgehog. He holds a Bachelor's Degree in Physics and Education from Case Western Reserve University and John Carroll University, and his current passions are cloud computing, Internet of Things, and Machine Learning and AI. His blog can be found at http://www.nicholaschase.com, or you can reach him on Twitter at @nickchase.

I would like to thank my wife Sarah Jane, who is also my Project Manager, for keeping me organized so that I could get everything (including the work on this book) done. I'd also like to thank the OPNFV Project for having faith in the idea behind this book, for all of the technical support, and for understanding the crazy schedule. Most of all, I'd like to thank Amar Kapadia for stepping in. This book wouldn't be possible without all of your hard work.

Understanding OPNFV

Table of Contents

What This Book Covers 10
Who This Book is For 12
Registration 12

1. What is NFV? **15**
NFV Business Drivers 15
What is NFV? 16
Benefits of NFV 18
NFV Use Cases 21
NFV at ETSI 22
NFV Requirements 24
Next Steps 27

2. NFV Transformation: Where's the CLI? **29**
NFV Drives Technology Changes 30
Organizational Impact 39
Starting the Journey 45

3. It's a bird, it's a plane, it's OPNFV! **47**
OPNFV Driving Factors 47
What is OPNFV? 48
Open Source Systems Integration 50
Benefits of Open Source 51
Multiple Approaches to NFV 52
So Why Should I Care Again? 53

    {Create-Compose-Deploy-Test}.Iterate   *53*

4. Upstream Projects in OPNFV   **55**
    NFV Management and Orchestration (MANO)   *56*
    Virtualized Infrastructure Manager (VIM)   *60*
    Software Defined Networking (SDN) Controller   *64*
    NFVI Compute   *67*
    NFVI Storage   *69*
    NFVI Hardware   *72*
    Out-of-Scope   *72*

5. OPNFV Upstream Contributions   **73**
    OPNFV Project Details   *74*
    Project Analytics   *79*
    Success Story: OPNFV Doctor   *80*

6. OPNFV Continuous Integration   **87**
    OPNFV RelEng: Release Engineering   *87*
    OPNFV Pharos: Community Lab Infrastructure   *91*
    OPNFV Octopus: Continuous Integration Project   *93*

7. OPNFV Automated Software Deployment   **97**
    What Are OPNFV Scenarios?   *97*
    Installers   *100*
    Additional Software Deployment Projects   *103*

8. OPNFV Continuous Testing   **107**
    Classification of Test Cases   *108*
    OPNFV Test projects   *110*
    Dashboards   *113*
    Plugfests   *115*

9. Writing VNFs for OPNFV   **117**
    Writing VNFs   *117*
    VNF Onboarding   *118*
    Clearwater vIMS on OPNFV   *121*

10. Utilize OPNFV to Drive Business   **127**

OPNFV and End-users	*127*
OPNFV and Technology Providers	*134*
OPNFV and Individuals	*134*
Getting Involved	*135*

11. Additional Resources — 137

Preface

Network functions virtualization, or NFV, is a once in a generation disruption that will completely transform how networks are built and operated. This mega-trend will affect telecom operators and technology providers alike. In fact, it's hard to imagine the promise of 5G being realized without NFV. This revolution will also reverberate across the non-telecom world as it transforms use cases ranging from corporations connecting branches to their central data center, to smart cities providing city wide connectivity, to manufacturers, utilities and connected car builders creating IOT networks.

Open source has revamped how enterprises build out their IT systems. Now, open source promises to do the same for NFV. OPNFV is the open source project for integrated testing of the full, next-generation networking stack to enable accelerated NFV. Engineering, network operations and business leaders in companies around the world are wondering what OPNFV is and how they can use it to their advantage. Whether you work for a telecom operator or a technology provider, this book will give you a complete overview of the OPNFV project so you can confidently provide direction to your technical team in terms of how to use or get involved with OPNFV.

What This Book Covers

Chapter 1, What is NFV?, provides an introduction to Network Functions Virtualization, the benefits of NFV, use cases, the role of ETSI, and NFV requirements.

Chapter 2, NFV Transformation: Where's the CLI?, covers model drive architectures, DevOps, cloud native software, the impact of NFV transformation on an organization, and tips on how to start the NFV journey.

Chapter 3, It's a bird, It's a Plane, It's OPNFV!, talks about the driving factors that were the impetus to create the OPNFV project, an overview of what OPNFV is, the concept of open source systems integration, the benefits of OPNFV, and finally, alternative approaches.

Chapter 4, Upstream Projects in OPNFV, provides a comprehensive view of all the NFVI, VIM and MANO projects integrated in OPNFV.

Chapter 5, OPNFV Upstream Contributions, briefly covers all the OPNFV "feature" projects that contribute requirements and code to the constituent upstream projects. We also take a deeper look at a successful project, OPNFV Doctor.

Chapter 6, OPNFV Continuous Integration, covers the infrastructure projects in OPNFV and takes a detailed look at the CI pipeline that fully automates integration, builds, deployment, and testing.

Chapter 7, OPNFV Automated Software Deployment, discusses the notion of scenarios, the four major installers that deploy these scenarios, and other deployment related projects.

Chapter 8, OPNFV Continuous Testing, talks about the various testing projects in OPNFV, test dashboards, and plugfests.

Chapter 9, Writing VNFs for OPNFV, provides a brief overview of VNF software architecture options and VNF onboarding, and ends with a concrete example of the Clearwater project onboarded by OPNFV for testing purposes.

Chapter 10, Utilize OPNFV to Drive Business, covers the all-important topic of "what's in it for me" (WIIFM) – in other words, how does OPNFV help end users, technology providers, and individuals? The chapter continues by providing specific ways you can get involved with OPNFV.

Chapter 11, Additional Resources, lists useful links for all 10 chapters.

Who This Book is For

This book is targeted at engineering, network operations, and business leaders who want an in-depth understanding of OPNFV and how it can help accelerate their NFV deployments. After reading this book, you will gain the confidence needed to guide your company's NFV transformation efforts, and to provide direction to your technology teams. The book is also useful for technology providers trying to understand how to leverage the OPNFV framework and for individuals looking to expand their career into the growing NFV ecosystem. The book assumes a basic level of technology awareness but does not require any hands-on knowledge of individual topics.

Reader Feedback & Questions

We welcome your feedback! Additionally, even though we have taken care to ensure the accuracy of the content, mistakes do occur. Please direct both the feedback and errata submissions to opnfvbook@mirantis.com.

Registration

As a valued book reader, register yourself on the Mirantis website to receive ongoing information about NFV related blogs, webinars, white papers and more! Register at https://content.mirantis.com/Understanding-OPNFV-eBook-Landing-Page.html and you will also get an electronic copy of this book.

Foreword

Since its founding in September 2014, the OPNFV project has been on an extraordinary journey. Watching our community embrace open source, execute multiple releases, and grow ever more mature in how we approach our upstream communities and our own activities has given me high confidence in our industry, our capabilities, and our commitment to transforming networking.

Working side-by-side with NFV end users gives us extraordinary insight into the real-world challenges of NFV deployments along with the many opportunities of evolving toward a software defined future. Through this ongoing and active collaboration, OPNFV facilitates the development and evolution of open source NFV across the industry via integrated testing of the next-generation networking stack. It's a formidable and inspiring challenge that is producing real business value to NFV end users, technology providers and individuals alike.

Participation in OPNFV is open to anyone, whether you are an employee of a member company or just passionate about network transformation. Reading this book will provide you with a solid foundation of open source NFV, the OPNFV project and community, as well as connect the dots on where you fit in and how you can benefit. The journey toward NFV is as much a cultural transformation of people and processes as it is any specific technology, and I welcome you to our vibrant, agile, and continuously improving community.

We hope you'll join us for the journey ahead.

Heather Kirksey
Director, OPNFV
April 2017

Understanding OPNFV

1
WHAT IS NFV?

NFV Business Drivers

Telcos, multiple-system operators (MSOs i.e. cable & satellite providers), and network providers are under pressure on several fronts, including:

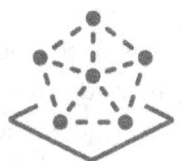

OTT/ Web 2.0

ARPU under pressure

Increased Agility

Over-the-top and Web services are exploding, requiring differentiated services and not just a 'pipe'.

Average revenue per user is under pressure due to rising acquisition costs, churn and competition.

Pressure to evolve existing services and introduce new services faster is increasing.

Enterprises with extensive branch connectivity or IOT deployments also face similar challenges. If telecom operators or enterprises were to build their networks from scratch today, they would likely build them as software-defined resources, similar to Google or Facebook's infrastructure. That is the premise of Network Functions Virtualization.

What is NFV?

In the beginning, there was proprietary hardware.

We've come a long way since the days of hundreds of wires connected to a single tower, but even when communications services were first computerized, it was usually with the help of purpose-built hardware such as switches, routers, firewalls, load balancers, mobile networking nodes and policy platforms. Advances in communications technology moved in tandem with hardware improvements, which was slow enough that there was time for new equipment to be developed and implemented, and for old equipment to be either removed or relegated to lesser roles. This situation applied to phone companies and internet service providers, of course, but it also applied to large enterprises that controlled their own IT infrastructure.

Today, due largely to the advent of mobile networking and cloud computing, heightened user demands in both consumer and enterprise networks have led to unpredictable ("anytime, anywhere") traffic patterns and a need for new services such as voice and video over portable devices. What's more, constant improvement in consumer devices and transmission technology continue to evolve these themes.

This need for agility led to the development of Software Defined Networking (SDN). SDN enables administrators to easily configure, provision, and control networks, subnets, and other networking architectures on demand and in a repeatable way over commodity hardware, rather than having to manually configure proprietary hardware. SDN also made it possible to provide "infrastructure as code," where configuration information and DevOps scripts can be subject to the same oversight and version control as other applications.

Of course, there was still the matter of those proprietary hardware boxes.

Getting rid of them wasn't as simple as deploying an SDN; they were there for a reason, and that reason usually had to do with performance or specialized functionality. But with advances in semiconductor performance and the ability of conventional compute hardware to perform sophisticated packet processing functions came the ability to virtualize and consolidate these specialized networking functions.

And so, Network Functions Virtualization (NFV) was born. NFV enables complex network functions to be performed on compute nodes in data centers. A network function performed on a compute node is called a Virtualized Network Function (VNF). So that VNFs can behave as a network, NFV also adds the mechanisms to determine how they can be chained together to provide control over traffic within a network.

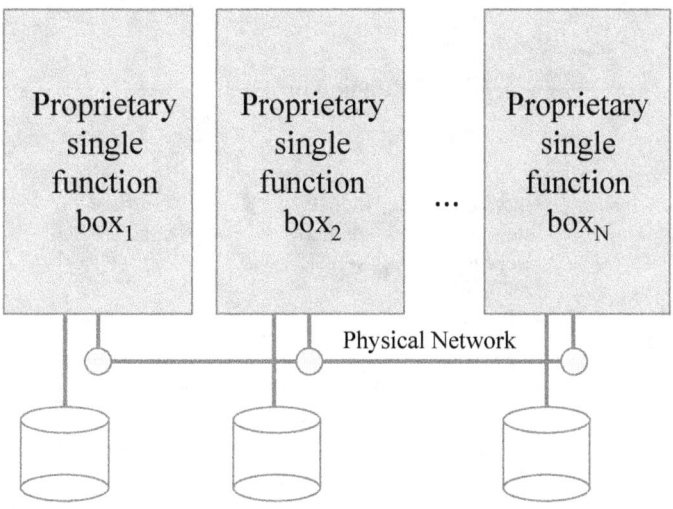

Simplified Network Architecture Before NFV

Understanding OPNFV

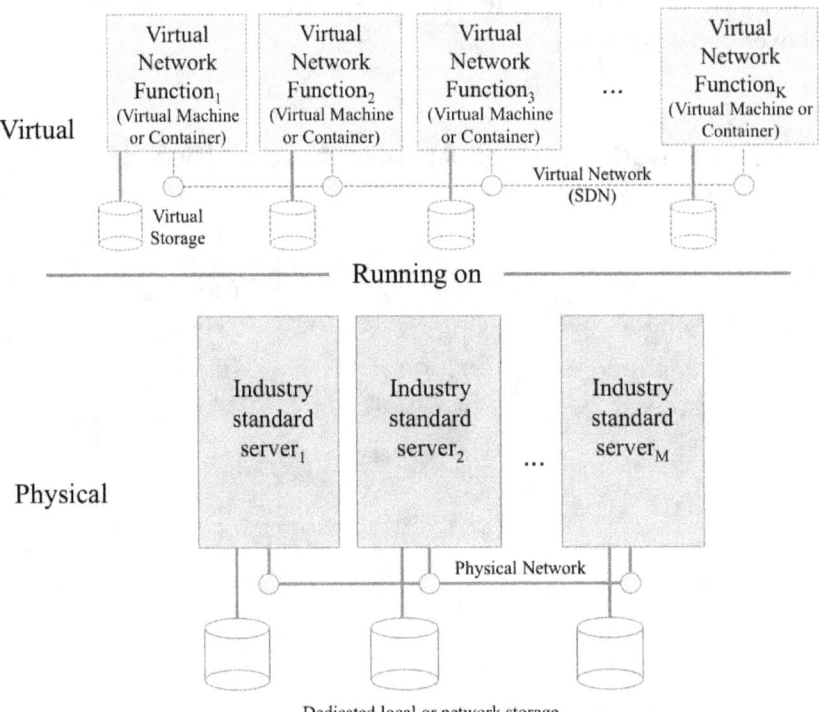

Although most people think of it in terms of telecommunications, NFV encompasses a broad set of use cases, from Role Based Access Control (RBAC) based on application or traffic type, to Content Delivery Networks (CDN) that manage content at the edges of the network (where it is often needed), to the more obvious telecom-related use cases such as Evolved Packet Core (EPC) and IP Multimedia System (IMS).

Benefits of NFV

NFV is based on the "Google infrastructure for everyone else" trend where large companies attempt to copy the best practices from the web giants to increase revenue and customer satisfaction while also slashing operational and capital costs. This explains the strong interest in NFV from both telcos and enterprises with numerous benefits:

Increased Revenue

New services can be rolled out faster (since we are writing and trying out code and vs. designing ASICs or new hardware systems), and existing services can be provisioned faster (again, software deployment vs. hardware purchases). For example, Telstra's PEN product was able to reduce the provisioning time for WAN-on-demand from three weeks to seconds, eliminate purchase orders and man-hours of work and reduce customer commitment times for the WAN link from one year to one hour.

Telstra's PEN Offering

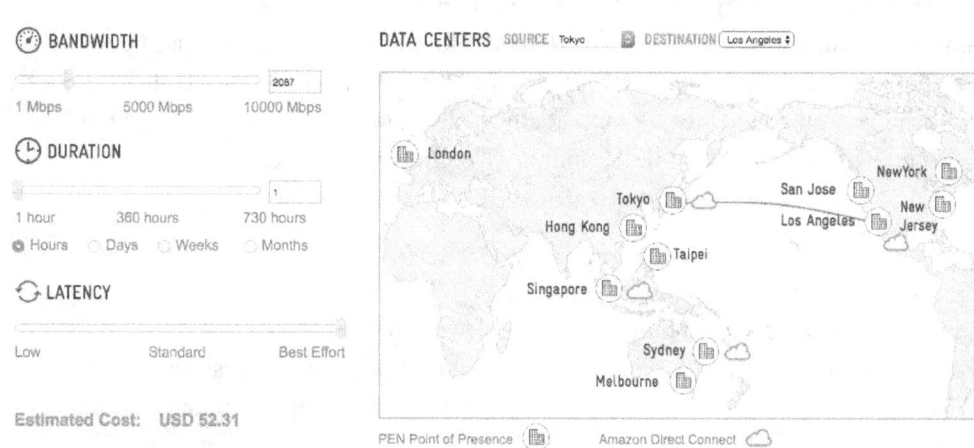

Improved Customer Satisfaction

With an agile infrastructure, no one service runs out of resources as each service is dynamically provisioned with the exact amount of infrastructure required based on the utilization at that specific point in time. (Of course, there's still a limit on the aggregate amount of infrastructure.) For example, no longer will mobile end users experience reduced speed or service degradation. Customer satisfaction also improves due to rapid self-service deployment of services, a richer catalog of services and the ability, if offered by the operator, to try-before-you-buy.

Reduced Operational Expenditure (Opex)

NFV obviates numerous manual tasks. Provisioning of underlying infrastructure, network functions and services can all be automated; even offered as self-service. This removes a whole range of truck rolls, program meetings, IT tickets, architecture discussions, and so on. At a non-telco user, cloud technologies have been able to reduce operations team sizes by up to 4x, freeing up individuals to focus on other higher-value tasks.

The standardization of hardware also slashes operational costs. Instead of managing thousands of unique inventory items, your team can now standardize on a few dozen. A bonus to reduced opex is reduced time-to-break-even. This occurs because, in addition to just virtualizing individual functions, NFV also allows complex services consisting of a collection of functions to be deployed rapidly, in an automated fashion. By shrinking the time and expense from customer request to revenue by instantly deploying services, the time-to-break-even can go down significantly for operators.

Reduced Capital Expenditure (Capex)

NFV dramatically improves hardware utilization. No longer do you waste unused cycles on proprietary fixed function boxes provisioned for peak load. Instead you can deploy services with the click of a button, and have them automatically scale-out or scale-in depending on utilization. In another non-telco industry example, a gaming IT company, G-Core, was able to double their hardware utilization by switching to a private cloud.

Using industry standard servers and open source software further reduces capex. Industry standard servers are manufactured in high volumes by multiple vendors resulting in attractive pricing. Open source software is also typically available from multiple vendors, and the competition drives down pricing. This is a win-win where reduced or elimination of vendor lock-in comes with reduced pricing.

Additionally, operators can reduce capex by utilizing different procurement models. Before NFV, the traditional model was to issue an RFP to Network Equipment Manufacturers (NEMs) and purchase a complete solution from one of them. With NFV, operators can now pick and choose different best-in-class vendors for different components of the stack. In fact, in some areas an operator could also choose to skip vendors entirely via the use of 100% open source software. (The last two option is not for the faint-of-heart, and we will explore the pros and cons of different procurement models in the next chapter.)

TIA Network's "The Virtualization Revolution: NFV Unleashed – Network of the Future Documentary, Part 6" states that the total opex plus capex benefit of an NFV-based architecture could be a cost reduction of up to 70%.

Freed up Resources for New Initiatives

If every operator resource is busy with keeping current services up and running, there aren't enough staff resources to work on new upcoming initiatives such as 5G and IoT. The side effect of reduced opex is that the organization will now have resources freed up to look at these important new initiatives, and so contribute to overall increased competitiveness. Or putting it another way, unless you fully automate the lower layers, there won't be enough time and focus on the OSS/BSS layer, which is the layer that improves competitiveness and generates revenue.

Example Total-cost-of-ownership (TCO) Analysis

Intel and the PA Consulting Group have created a comprehensive TCO analysis tool for the vCPE use case (see below). In one representative study conducted with British Telecom, the tool was populated with assumptions for an enterprise customer where physical network functions from the customer's premise were moved to the operator's cloud. In this study, the tool shows that the operator can reduce their total cost by 32% to 39%. The figure encompassed all costs including hardware, software, data center, staff and communication costs. The TCO analysis was conducted over a five-year period, and included a range of functions such as firewall, router, CGNAT, SBC, VPN and WAN optimization. These results are representative and will obviously change if another study has different assumptions. Also, as mentioned earlier, cost is only one of the many benefits of NFV.

> **One study found 32%-39% TCO reduction for vCPE**

NFV Use Cases

Since the initial group of companies that popularized NFV was made up primarily of telecommunications carriers, it is perhaps no surprise that most of the original use cases are related to that field. As we've discussed, NFV use cases span a broader set of industries. Instead of covering all use cases comprehensively, we are going to touch upon the three most common:

vCPE (Virtual Customer Premise Equipment)

vCPE virtualizes the set of boxes (such as firewall, router, VPN, NAT, DHCP, IPS/ IDS, PBX, transcoders, WAN optimization and so on) used to connect a business or consumer to the internet, or branch offices to the main office. By virtualizing these functions, operators and enterprises can deploy services rapidly to increase revenue and cut cost by eliminating truck rolls and lengthy manual processes. vCPE also provides an early glimpse into distributed computing where functionality in a centralized cloud can be supplemented with edge compute.

vEPC (Virtual Evolved Packet Core)

Both the sheer amount of traffic and the number of subscribers using data services has continued to grow as we have moved from 2G to 4G/LTE, with 5G around the corner. vEPC enables mobile network operators (MVNO) and enablers (MVNE) to use a virtual infrastructure to host voice and data services rather than using an infrastructure built with physical functions. A prerequisite to providing multiple services simultaneously requires "network slicing" or the network multi-tenancy, a capability also enabled by vEPC. In summary, vEPC can cut opex and capex while speeding up delivery and enabling on-demand scalability.

vIMS (Virtual IP Multimedia System)

OTT competitors are driving traditional telco, cable and satellite providers towards offering voice, video, and messaging over IP as a response. A virtualized system can offer the agility and scalability required to make IMS an economically viable offering to effectively compete with startups.

This list is by no means comprehensive, even in the short term. Numerous other use cases exist today and new ones are likely to emerge. The most obvious one is 5G. With 50x higher speeds, 10x lower latencies, machine-to-machine communication, connected cars, smart cities, e-health, IOT and the emergence of mobile edge computing and network slicing, it is hard to imagine telecom providers or enterprises being successful with physical network functions.

NFV at ETSI

In 2012, ETSI took the reins of the standardization process for NFV, with the goal of defining just what NFV was and what the industry was going to do. The Industry Specification Group (ISG) for NFV at ETSI has grown from seven to more than 300 individual companies, and has published more than 40 documents with details of how NFV should be implemented and tested. The ongoing contribution of ETSI in furthering the cause of NFV in the standards arena has been invaluable. Let us briefly look at three important NFV-related documents created by the ISG.

NFV Architectural Framework Document

This document is important because it forms the reference against which all the ISG working groups work.

ETSI NFV Architecture (source ETSI)

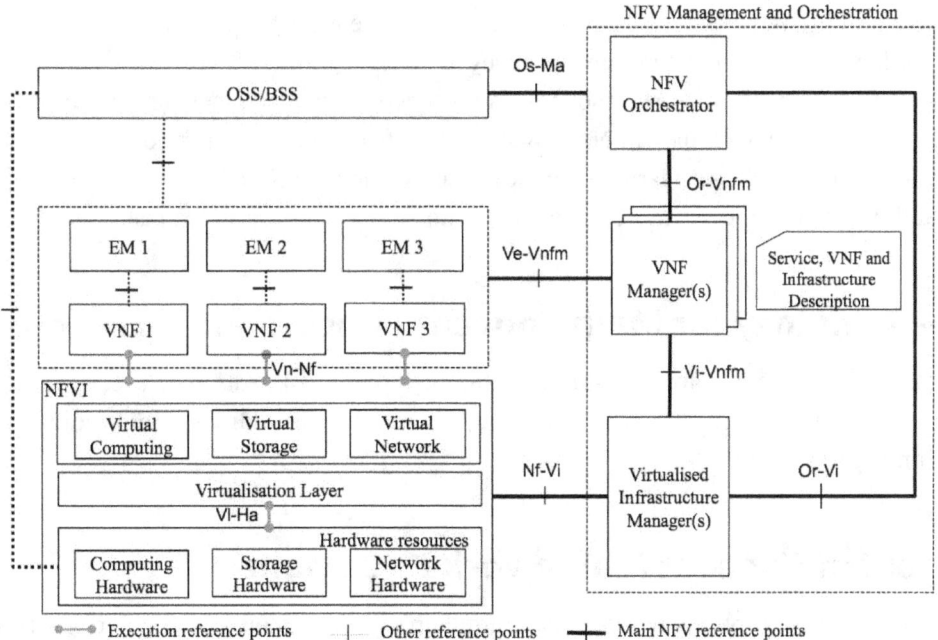

The actual NFV Architectural Framework consists of several components:

- The Network Functions Virtualization Infrastructure (NFVI) is the actual infrastructure on which these services are delivered. NFVI includes:
    - Physical hardware such as servers, storage, network hardware, and
    - Virtual infrastructure such as virtual compute (virtual machines or containerized virtual operating systems), virtual storage, and virtual networks (software defined networking).
- One or more Virtual Network Function (VNF) runs on top of the NFVI, providing the actual software-based versions of network functions. In most cases, these are functions that have replaced traditionally hardware-implemented functions, such as switches, firewalls, and load balancers. This software functionality should be hardware-independent, but often requires hardware optimizations for performance reasons.
- Both the NFVI and the VNFs must be managed, of course, and this is handled by NFV Management and Orchestration (MANO). MANO can be complex, as it must both provision and manage the lifecycle of both software and hardware resources as necessary. NFV MANO usually doesn't act in isolation, interacting with the existing external OSS and BSS landscape.
- Finally, NFV MANO works with metadata that describes services, VNFs, and

infrastructure requirements.

The Architectural Framework is important because to enable multiple players to determine different portions of the architecture in which they're going to work, it's necessary to abstract and virtualize the various components so they can be decoupled. Once they have been decoupled, the ways in which they interface with each other can be determined and standardized. Even with the standardization, however, the actual task of establishing interoperability between the various components is an ongoing and complex task.

NFV Terminology for Main Concepts Document

This document is a repository for terms used within the other documents. The ISG hopes both that other groups will use their terms, and that they will contribute additional terminology to simplify conversations about NFV-related issues.

NFV Proof of Concept Framework Document

All of this theory means nothing without actual implementations to prove that it works. To that end, the ISG has also produced guidelines for those who wish to create Proofs of Concept (PoCs) based on the work it's produced. These guidelines include a proposal template, as well as requirements for who must participate. For example, a PoC must include at least two vendors and at least one network operator/service provider who is also a member of the NFV ISG.

The purpose of these PoCs is to provide feedback about one or more interoperability or technical challenges that must be overcome for a fully-functional NFV ecosystem.

NFV Requirements

While it is easy to fixate on the positive aspects of NFV, there are also several requirements that must be solved before NFV can succeed. Some of the more important "carrier-grade" requirements are discussed below.

Security

Security is vital in today's world, where new threats are discovered every day, and cyber-espionage is a pervasive problem.

That said, it's important to understand that the nature of virtualized infrastructure does bring with it additional security concerns that come with NFV. The NFV ISG Security Expert Group looked at the situation and came up with a list of 10 security concerns that are particular to virtualized environments:

- **Topology validation and enforcement:** Topology validation and enforcement is the process of making sure that traffic can only go where it's supposed to. In a physical network this is more straightforward; traffic can be administratively separated. In a virtualized environment, this is more involved – the theoretical possibility exists for traffic to go anywhere, and for inadvertent loops to be exploited.
- **Availability of management support infrastructure:** In this case, we're talking not about the higher-ups in the organization, but rather about the ability to administer a system after it has crashed. While the preference is for a separate physical network, for cost reasons, this is normally provided by connectivity through a separate port; it's important to make sure that this access is protected, at a minimum by VPN.
- **Secured boot:** It's important that a host be able to verify that VNFs are genuine before firing them up, and for a VNF to know that a host hasn't been compromised before performing sensitive operations.
- **Secure crash:** When and if a VNF crashes, the system must make sure to destroy all references to it, such as open connections or data that's been persisted to the file system.
- **Performance isolation:** In a public cloud, this is often called the "noisy neighbor" problem; two processes running on the same core or sharing other resources can interfere with each other's performance. To solve this problem, you can specify that processes must have their own reserved resources, but that brings its own challenges based on resource usage.
- **User/tenant authentication, authorization and accounting:** NFV increases the number of layers at which users must be identified, as well as the complexity of accounting requirements. Some security risks can be mitigated with the use of tokens rather than direct identity passing, but overall, NFV does present a larger attack surface than traditional hardware environments.
- **Authenticated time service:** In this case, it's a matter of a potential for attacks, rather than actual attacks that is being pointed out; some code relies on timing to function properly, so it queries multiple sources to get the "true" time. A VNF, however, can only talk to a hypervisor, so there is no possibility of getting multiple "votes". (Note that this is actually a cloud issue, as opposed to a NFV-specific issue.)
- **Private keys within cloned images:** In an ideal world, cloned images would not include private keys or passwords for access, and instead that information would be injected at boot time. Unfortunately, we do not live in an ideal world. This is a problem that's not necessarily a technical challenge, but an operational one.
- **Back-doors via virtualized test & monitoring functions:** Another operational issue involves test or debug interfaces that may be enabled on an official or unofficial basis to enable operators to test or debug production components. If you must leave them in

place, they must be protected from attack.
- **Multi-administrator isolation:** Mitigation of Multi-Administrator Isolation issues is an area that still requires research. It involves the technical consequences that arise when the administrator of a VNF is meant to heave more secure access than the administrator of the overall system. (This situation comes up in cases of "lawful interception.") The problem is that in general, it's difficult to secure resources on a system from the system's administrator.

The NFV Security Expert Group has presented these issues to the NFV ISG, which is working with other standards bodies to try and alleviate these problems.

Performance

NFV infrastructure is designed to emulate a hardware environment with one made up of software, but that doesn't mean that you can just drop the software into your cloud and go. Most of what NFV does requires extremely performant systems – but they're required in an environment where the VNF doesn't always know what it's getting.

What this means is that it's up to the operator to configure and deploy the software in such a way as to take advantage of best practices when it comes to performance. In fact, according to the NFV ISG, the difference in speed between using NFV and using NFV correctly can be as much as 1000%. Needless to say, NFV's promise of reducing capital expenditures can quickly get eroded if instead of one proprietary box, you have to buy multiple industry standard servers that perform poorly. In short, as an industry, we need to close the gap between NFVI and physical network function price-performance.

Interoperability

NFV is meant to be an ecosystem that is occupied by multiple players, so it's especially important that all their solutions can work together. To that end, it's crucial to have specifications that define interoperability between the various portions of the overall system, such as the Orchestrator, the Virtualized Infrastructure Manager (VIM), the Virtual Network Function Manager (VNFM), the VNF, and the actual Network Functions Virtualization Infrastructure (NFVI). Also, interoperability with legacy OSS/BSS systems is just as important.

In this case, both the open source approach and the encouragement of PoCs can help reach this goal, as multiple operators attempt to compose systems out of multiple components.

Easy to Operate

Amazon has an army of cloud specialists to operate Amazon Web Services. If we require the same of NFV deployments, we will either increase the operational cost dramatically and/or decelerate its deployment. Therefore, a key requirement for NFV to succeed is operational simplicity. From "day-1" initial deployment to "day-2" post-deployment operations such as changing configurations, adding capacity, adding functionality, modifying reference architectures, updates and upgrades, all need to be simplified dramatically for NFV to be deployed at a large scale. And by the way, these operations need to be performed without interrupting existing VNF workloads. Additionally, log monitoring, metrics and alerts need to be provided for an entire cluster in aggregate, in a simple to consume manner.

NFV-Specific Requirements

NFV has numerous unique requirements that are distinct from, say, enterprise requirements. Let us look at a couple of the most important ones:

- **Service assurance:** Arguably the availability demands on a VNF are higher than a web or mobile cloud application. After all, an OnStar customer relying on the LTE network to request emergency services in response to an accident cannot tolerate any loss of availability of the network! So how are we to detect the root cause of a failure and address it efficiently? This is the crux of service assurance, where we want to be able to trace the root cause of a failure to the actual infrastructure fault and then address it in the most optimal manner. Over time, we want to predict potential failures before they happen, and be proactive rather than reactive. The holy grail is to use artificial intelligence and machine learning to predict and respond to issues.
- **Service function chaining:** This is a concept where a service provider needs to apply services, consisting of multiple chained VNFs, to data traffic between two endpoints. These services could be anything from authentication to security to transcoding. What makes the requirement even more interesting is that the insertion of VNFs is policy-based and dynamic. So, two users requesting a video from the same website could have different VNFs inserted in the middle depending on the device they are using, their payment plan, and, say, their location.

Next Steps

NFV is gaining momentum both in the telecom and enterprise fields, which is leading to new service models and use cases, as evidenced by a study conducted by Heavy Reading in August, 2016, where 98% of the service providers surveyed indicated that NFV is part of their strategy. More generally speaking, NFV is nothing but a network-heavy workload (as opposed to compute-heavy). With the rise of IOT, machine-to-machine communication, intelligent edge devices (wearables, drones, augmented reality, and so on) NFV will see widespread adoption in enterprises as well. Net-net, everyone needs higher quality, more dynamic, increasingly elastic, and better tuned connectivity to their applications and services.

In addition to ETSI, numerous standards and open source organizations have started working on NFV. A leading organization in this effort is the Linux Foundation. And the Linux Foundation's OPNFV project. In the next chapter, we'll look a slight detour to look at what an NFV-transformation means for an organization, before coming back to OPNFV in Chapter 3.

2

NFV TRANSFORMATION: WHERE'S THE CLI?

We have already established that network and software development agility and the ability to out-innovate competitors will become increasingly important, and that NFV is a way to achieve this agility. However, to be successful with NFV, users will need to undergo a transformation that is broader than just technology. In fact, NFV affects a variety of non-technology related aspects such as processes, organizational structure and skill set acquisition. NFV also affects secondary factors, such as business models and procurement methodology. Let's first review three major technology changes NFV requires, and then look at the resulting organizational impact.

Agility is Now *the* Key Differentiator

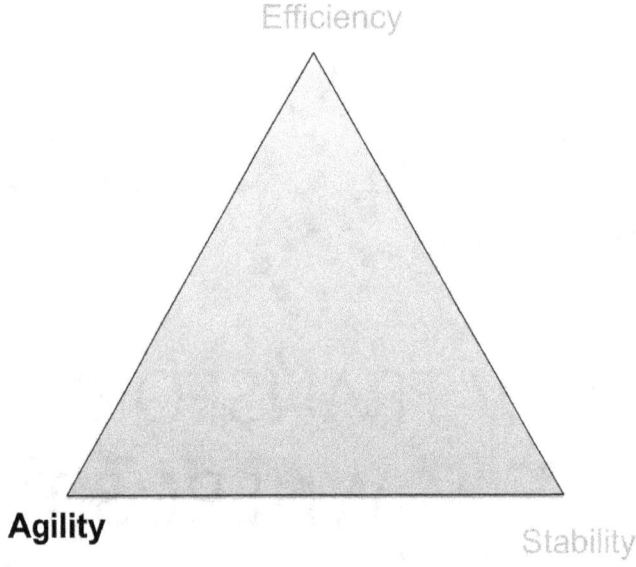

NFV Drives Technology Changes

To be successful, operators will need to embrace three broad technology changes:

- Model driven architectures
- DevOps
- Cloud native VNFs

Most of these changes are still in their infancy, but as they mature they'll become increasingly important:

Model Driven Architectures

It is not uncommon for a new service deployment with physical network functions to take 30-60 days. Equipment must be purchased, configured and deployed, often using vendor-specific proprietary syntax, tooling and professional services. Then these physical boxes have to be connected to form a service.

By replacing these fixed function boxes with industry standard servers allows the infrastructure to be provisioned rapidly, but still does not solve the problem of deploying a service rapidly. Even with virtualization, VNFs have to be installed and configured, and then connected to form a complete service. Bringing up the service often has dependencies whereby functions have to be provisioned in a certain order. Finally, the monitoring infrastructure for each VNF must be put in place. Once the service is up and running, "day-2" activities kick in in terms of updates, upgrades, scaling-out, scaling-in, self-healing, configuration changes, and so on (see *Full Automation of NFV Software Lifecycle* figure below). Doing this manually can take weeks or months of effort.

That's where model driven architecture comes into the picture. Model driven templates describe the entire network service, with configurations for each VNF, its topology and connectivity, various dependencies and policies for triggering actions specified in a human-readable file. These templates can also capture management, monitoring and capability APIs. With the entire network service descriptor (NSD) available as code, there are no manual steps or tickets to provision services, and the service is created, composed and deployed in exactly the same manner, every time. This is true even across different environments ranging from dev/test to production, even with different underlying NFVI, VIM or SDN controller elements. Moreover, model driven templates can also be put in revision control for better tracking and compliance.

Model Driven Architecture Fuels NFV

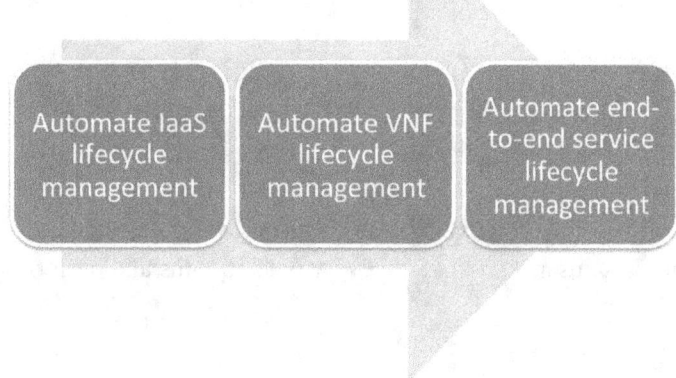

If model driven templates are so useful, why stop with just network service descriptions? In fact, these templates are used for lower level components such as VNF descriptors, software application and infrastructure resources, and higher level items such as products (collections of services with commercial attributes) and offers (bundles of products with marketing configurations). While lower level templates concentrate on the resource, policy, topology, configurations and dependencies, higher-level templates include process, order, pricing and business rules. To sum it up, a model driven architecture vastly reduces the time it takes operators to deliver new products and services to their customers, and enables rapid onboarding of 3rd party solutions.

DevOps

The second area of technology change to required ensure a successful NFV transformation is DevOps.

Every layer of the NFV stack (VNFs, NFVI, and MANO) is seeing innovation at a rate faster than anything witnessed in the prior era of physical network functions, and this pace is not going to slow down. To absorb this rate of innovation, organizations will have to fundamentally change their approach to technology components.

Continuous Innovation Requires a New Approach

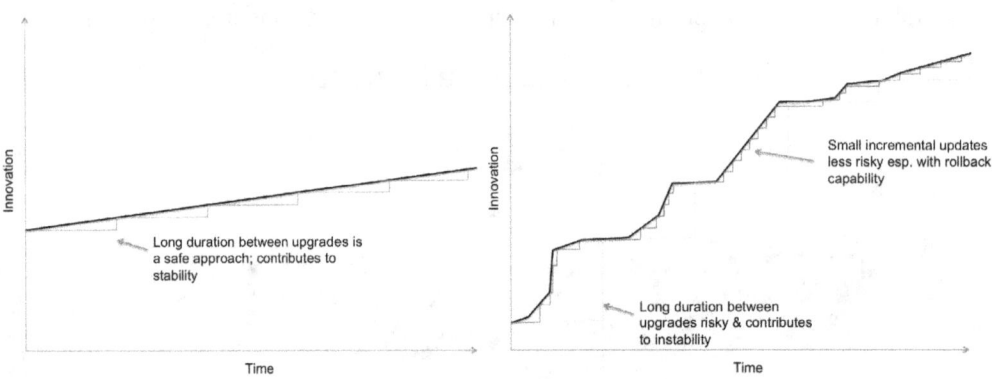

The figure on the left shows, on a relative basis, slower innovation created by vendors at a predictable rate. In this case, users can upgrade every 6-18 months and perform routine operations in the interim.

The right-hand side figure shows how that same safe and stable approach breaks down with rapid, continuous, and unpredictable rates of innovation coming from not just vendors but increasingly from open source communities. A 6-18-month upgrade is now going to create massive risk to an entire service or class of services.

What's more, this window also keeps innovation away from the end users' hands for an unnecessary 6-18 months, just as it keeps valuable user feedback away from developers. What makes sense instead is small incremental changes, made regularly.

What does that kind of environment look like? Let's take Netflix as an example. Netflix isn't strictly an NFV architecture, but it does exemplify this type of development; Netflix makes thousands of small changes to their production environment every single day. Using the old method of "forklift upgrades", that kind of pace would have been not just unheard of – it would be suicide.

This process of making continuous incremental changes results from a practice called DevOps. The term "DevOps" has become slightly overused lately, leading to some argument over what it actually means (solid Wikipedia definition notwithstanding). Some view it as a mindset where development and ops teams work in tight collaboration. Others view it as a set of activities. In reality it is perhaps a bit of both.

Using the former viewpoint, DevOps is about two groups with different goals coming together. Development's[1] goal is to increase software velocity. Clearly, development is incentivized to create change. Operations, on the other hand, is measured in terms of network availability, Mean Time Between Failures (MTBF), Mean Time to Repair (MTTR) and other operational metrics. So, while Operations, isn't necessarily trying to prevent change, they are obligated to operationalize change and make sure that it happens without disruptions. DevOps is a way to encourage collaboration by breaking down the walls between these two teams and helping them embrace change without jeopardizing stability.

If we view DevOps from the second lens of "activities", it involves these following steps: continuous integration, continuous testing, continuous delivery, continuous monitoring.

---

[1] In the case of NFV, the development of the NFV software stack and VNFs is typically not done in-house. For that reason, "Dev" in DevOps for NFV may be loosely considered external vendors and open source communities. This is different from web giants, where most of the development is in-house.

Steps Involved in DevOps

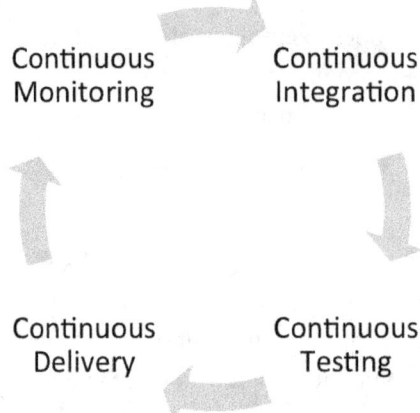

Classic DevOps for internally developed software components is different from DevOps for NFV, where software components are sourced from vendors and open source communities. In this sense, NFV users do not own the development-integration-test cycle of VNFs, instead acting as integrators that compose these third-party VNFs. Another contrast is that classic DevOps assumes homogeneous services, while DevOps for NFV is all about diversity. For these reasons, DevOps for NFV is distinct and different from classic DevOps. Let's take a further look at DevOps for NFV:

- **Continuous integration (CI):** The idea here is to continuously build and integrate code. When the development is in-house, this could mean that literally every developer commit results in a new integration. However, with third party and open source software, this typically means an integration effort can be kicked off every time there is a release of some external software. With so many components (hypervisor, VIM, MANO, SDN controller, VNFs), you can expect there to come a point in the future where you will see some change just about every week, if not every day!
- **Continuous testing:** It is not enough to just integrate the software. Even though each piece of software has been tested by itself, the entire stack has to be successfully integrated and tested in a development or staging environment. Naturally, all this testing must be automated. It is not feasible to run manual tests at such a high frequency.
- **Continuous delivery (CD):** Once the software passes all automated tests, it needs to be deployed into production. Ideally, the new software component co-exists with the older version, and the new component is made available only to a subset of the affected services. These techniques are called canary or blue-green deployments. If the new component holds up in its initial limited production deployment, it's expanded to the rest of the production environment and the old one can be retired. If not, you can simply rollback the changes. In this way, small incremental changes promote stability. They

isolate failures to just one component at a time. So, you are unlikely to have an entire service or set of services fail, which is a distinct possibility with a forklift upgrade.
- **Continuous monitoring:** The final piece of the puzzle is to monitor NFV services continuously. Every log, metric, and alert has to be processed by automated big data systems, because lifecycle management decisions for any NFV software component must be made automatically based on policies. It is simply not possible to have a human being monitor all individual components. That's why you should never see CLI access to an individual VNF. The concept no longer makes sense in NFV!

To achieve DevOps successfully, organizations require automated tooling. CI (including continuous testing), and CD are often automated as a pipeline. In chapter 6, we will look at the OPNFV CI pipeline in more detail. In general, a CI/CD pipeline, followed by continuous monitoring, could be generalized as follows:

Representative CI/CD Pipeline and Ongoing Monitoring

Again, don't worry if this topic feels completely futuristic, we don't anticipate a full DevOps for NFV approach to take hold for at least a few years.

Cloud Native Software

The third area of technology change to ensure a successful NFV transformation is cloud native software. The entire NFV stack needs to, over time, migrate to a cloud native software architecture. Analogous to DevOps, cloud native NFV is similar to, but not the same as cloud native apps for consumer and enterprise use.

Simply put, cloud native NFV software architecture requires the following:
- Ability to scale out (to multiple nodes), scale in, and self-heal in an automated fashion.
- Ability to withstand, even thrive in an environment where hardware failures are commonplace.
- Support the DevOps for NFV approach (see above section) to fully automate the entire lifecycle of every component of the NFV software stack.

In short, every step below has to be software defined, with zero manual steps:

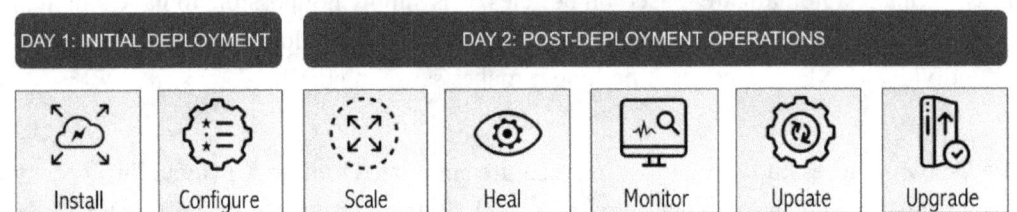

These three requirements necessitate re-architecting software applications around a set of core principles. The list below is not supposed to be complete; modern cloud native software is based on a principle called "twelve factor applications," but while you'd think that NFV was a shoo-in for "cloud native" computing, it turns out there are some differences.

Twelve-factor apps, you see, are meant to stateless – they're all about separating processing and state management so that any part of a system is the same as any other part. However, networking applications are all about state management, so the normal cloud native assumptions need to be translated for NFV[2]:

- **Microservices with API as contracts:** A VNF needs to be split into small pieces or microservices that are loosely coupled and communicate through well-defined APIs. Both the code and the data model has to be split. This approach provides benefits to both operators and VNF vendors. Operators benefit by being able to scale VNFs in a very granular fashion, reduce the impact of faults (by limiting the blast radius to one microservice) and manage the lifecycle of VNFs using a DevOps mechanism. Vendors benefit by being to roll out new features faster and eliminate dead features. As we have discussed, VNFs are different from classic apps, in that they have a datapath that performs packet processing and a control-path that provides management, signaling etc. For this reason, microservices have to be used with care. The datapath is not an obvious candidate, as multiple microservices in the datapath could create inefficiencies due to multiple packet hops, each with switching and packet processing overhead.
- **Decoupled from hardware:** Most benefits from the DevOps section can only be

---

[2] We mostly talk about VNFs, but the principles apply to other components of the NFV stack as well

experienced if the workload is completely decoupled from the underlying hardware and can request resources through APIs. This, of course, is achieved through virtualization. Decoupling doesn't mean divorcing a VNF from hardware, since in NFV there will always be some VNFs that need servers with specific features, for instance data plane acceleration. Over time, new paradigms such as Platform-as-as-Service (PaaS) and "serverless apps" (where the code is executed in response to events while completely abstracting the underlying server, operating system and related infrastructure) will further promote VNF-hardware decoupling.

An additional dependence on hardware is that traditional apps expect availability to be an attribute of the underlying infrastructure. So, it is the infrastructure that has to guarantee the required 9's of availability. In a cloud native application, the burden shifts to the VNF. The hardware is assumed to fail all the time. The current paradigm is often referred to as 'Pets' (i.e. each node has to be cared for) vs. the new paradigm that is referred to as 'Cattle' (i.e. instances are dealt with as an aggregate, and individual instances may come and go).

Example 1: Old Contract Between VNF & 'Pets' Infrastructure

- CPU brand, type, frequency, number of sockets
- Memory size
- NVMe flash brand, size
- Hard drive brand, type, speed, size, quantity
- NIC brand, performance and quantity
- Power supply and fan specifications and quantity

Example 2: New Contract Between VNF & 'Cattle' Infrastructure

- vCPU quantity
- Memory size
- HDD and/or SDD block storage size
- I/O with specific performance characteristics
- Data plane acceleration or other special features as required

- **State isolation:** In classic cloud native applications, the state for the application is isolated to database (SQL or NoSQL) services. All other functionality is in stateless services. Again, since networking is essentially all about distributed state management, state isolation applies only within the context of a microservices in VNFs. Each microservice should separate transaction processing logic from its state. This promotes

faster software development velocity and the ability to standardize on persistent database or storage layers. The database or storage layers could span the entire range of performance from memory, flash, block storage to object storage.
- **Scale-out vs. scale-up:** Cloud native apps scale by increasing the number of instances rather than adding more resources to a given instance. This is critical in terms of achieving hardware independence and performance that scales-out or in, in response to utilization. A corollary to scale-out is self-healing, where the system responds by creating new instances of a microservice if existing instances become unavailable.
- **Antifragility:** This is probably the most important yet neglected factor. An antifragile system is better than stable. It is a system that improves in stability as stress is applied to it in the form of infrastructure and other failures. Networking services have been designed to be antifragile since the Internet was first invented, but this is new for VNFs. In fact, Netflix has pioneered a new field of "chaos engineering" where they fail instances, networks, availability zones and entire regions to test to ensure antifragility. In production. All the time.

In reality, it takes a long time to get to all of an organization's workloads to be cloud native. It took Netflix 7 years to completely move from traditional applications to cloud native! For most users, in the interim, many of the benefits of NFV can be realized even with cloud hosted VNFs (traditional VNFs virtualized) or cloud optimized VNFs (partially cloud native VNFs). In fact, most VNF vendors are taking this exact approach by simply virtualizing their physical network functions and doing some basic performance optimizations to begin with.

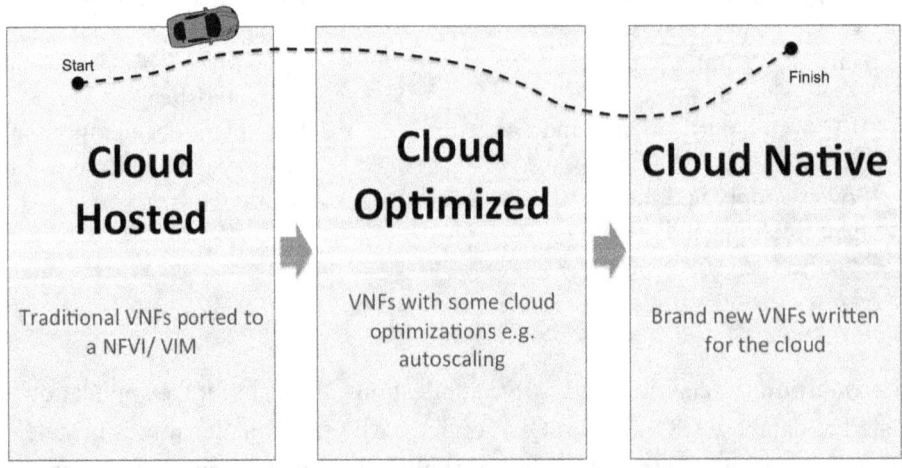

Organizational Impact

The three above changes – model driven architectures, DevOps and cloud native VNFs – impact many aspects of your organization, requiring a transformation. A complete discussion of what that transformation might look like is outside the scope of this book. Instead we will raise questions to help you determine the readiness of your organization for NFV. Score each question on a scale of 1 through 5 (1=not ready, 5=fully ready). For every section, the scores mean the following:

Score	Readiness
20-25	Completely ready
15-20	Almost ready
10-15	Somewhat ready
5-10	Significant transformation required
0-5	Transformation not yet begun

Impact to the Organization Structure

Technology teams within telecom operators are often organized by physical network functions. Furthermore, IT, network and OSS/BSS teams are segregated. For NFV, the organizational structure may need to be reorganized around a common pool of infrastructure and workloads. And skills from all the above functions need to be pooled together as well.

Moreover, cloud native thinking might force even more dramatic changes. For example, Twitter assigns one team to one microservice. The team is no more than seven individuals. Each team is cross-functional and owns the complete job of developing, testing, deploying and monitoring their microservice in production. Teams are typically collocated, and run like a startup. They make their own technology decisions, prioritize their own backlog and are held accountable as a team. There is of course cross-team coordination and negotiation. This frees up the platform team; it is only responsible for operating the actual hardware and providing automation tools. Failures are inevitable, and when there are failures, there is a blameless post-mortem across multiple teams.

Organizational structure readiness questionnaire:

Question	Readiness: 1-5 (1 = Not ready, 5 = Completely ready)
Has there been any example where cross-functional groups formed teams for new technology?	
Can you imagine completely autonomous teams per NFV service or component (such as NFVI, VIM, MANO) that owns dev, test, release *and* production?	
Can you imagine a decentralized decision-making approach where each team, within reason, could make their own technical decisions?	
Can you imagine the communication between NFV services teams and the platform team to be completely automated?	
Are there mechanisms to discuss requirements that affect multiple teams, and to coordinate these requirements?	

Impact to Process

During the 2016 OpenStack Summit in Austin, Jonathan Bryce talked about a customer who took 44 days to deploy an application. Once the company implemented OpenStack, they were able to cut this down to a disappointing 42 days. They had left all the checkpoints and manual signoffs intact! Once the cultural issues and processes were addressed, the duration was cut to 2 hours. This proves that to succeed, organizations need "software defined people".

Processes may range from the mundane, such as checkpoints and signoffs, to more complex, such as inventory management, security, service assurance, auditing, and so on, which are all difficult in a virtual world as instances spin-up and down. Often there is a lot of bureaucracy in place for any change, put in place to promote stability and reduce risk. With NFV, a complete review of all processes will be required, and those that do not truly contribute to stability will have to be replaced based on a culture of trust and accountability. Any spreadsheet, meeting, ticket, or handoff that doesn't truly add value must be eliminated. To goal is to move away from processes organized around physical network functions to a cloud-based approach.

Process readiness questionnaire:

Question	Readiness: 1-5 (1 = Not ready, 5 = Completely ready)
Can you imagine an intern pushing an update to a service in production on their first day at work? (Another imperfect analogy, but in the web world Etsy *requires* their interns to push a change to their live site on their first day at work)	
Could you imagine an engineer experimenting with a new feature on the production network on 1% of your user-base without seeking any permission? (Twitter allows this.)	
Do you use Agile methodology? Is the methodology truly Agile or is it waterscrumfall (that is, a hybrid of waterfall and Agile)?	
Can your finance team approve hardware purchases not allocated to a specific cost center, but for an aggregate NFV cloud?	
Can your compliance, asset inventory, and security teams deal with a dynamic virtual infrastructure?	

Impact to Technology

The impact of technology is perhaps the clearest. The organization must be able to absorb these new technologies and use them effectively.

Technology readiness questionnaire:

Question	Readiness: 1-5 (1 = Not ready, 5 = Completely ready)
Do you have a plan for what NFV software needs to be cloud	

hosted, cloud optimized or cloud native?	
Would your organization be receptive to the principles of Chaos Engineering?	
Is your organization ready to set up DevOps pipelines and automated testing?	
Can your operations team adapt to model driven architectures?	
Can you imagine a monitoring framework where you cannot access single VNFs or instances, but rather the management is done on the aggregate cloud?	

Impact to Skill Set Acquisition

The technology stack in NFV requires a higher level of skill set than with physical functions. Teams managing NFV deployments will need skills in both functional areas, such as understanding new components such as NFVI, VIM, SDN controller, MANO, and so on. And non-functional areas, such as lifecycle management of infrastructure and VNFs, monitoring, troubleshooting, and so on. These skills will have to be inculcated organically, because there just aren't enough external people you can hire with these skills. When available, these resources may also be more expensive.

Skill set readiness questionnaire:

Question	Readiness: 1-5 (1 = Not ready, 5 = Completely ready)
Are employees permitted any time and/or access to lab resources to learn new things on their own?	
Can you imagine a company-wide Plugfest/Hackfest to promote intense collaboration around activities such as tool development, interop testing, VNF onboarding and operational best practices?	
Are there formal training programs? How easy is it for	

technologists at your company to attend relevant trade shows or summits?	
Are there internal workshops or knowledge transfer programs? Will teams share learnings with each other?	
Are there post-mortems on projects? Can you imagine an environment where upper management will not assign blame during these post-mortems?	

While organizational structure, processes, technology and skill set acquisition are the primary areas impacted, there are also two secondary areas that need to transform: business models and procurement. Unlike the primary areas, we will skip the questionnaire for secondary areas.

Secondary Impact: Business Models

Current telecom operator business models have been optimized for mass market. With new trends such as IoT and network slicing, business models will have to be tailored for specific industries and business cases. One size fits all business models are not going to work. If the customer transaction will be better suited to different durations, such as daily, monthly, or even hourly, and different metrics such as number of cities visited using virtual reality, or number of Pokemons captured, and so on, then the business models will have to be adapted accordingly.

Secondary Impact: Procurement

How your organization wants to deal with vendors will need to be re-evaluated with NFV. A variety of new models open up.

Procurement Models

DIY Software	Vendor Software	Vendor Software	Vendor Software
DIY Integration	DIY Integration	System Integrator	System Integrator
DIY Management	DIY Management	DIY Management	Managed Services

There are many sub-variations, but broadly speaking the models are:

- **100% do-it-yourself (DIY):** In this model, the user purchases commodity hardware and utilizes only open source software. This results in staffing and, more importantly, retention of in-house software engineering, integration and operations. A substantial investment is required to create architecture, engineering, testing, DevOps, integration, interop testing, bug fixing teams and ops support as well. Furthermore, operational teams are required for design, deployment, monitoring, lifecycle management, break/fix support, and so on. In addition to investment and a level of sophistication, the user would also need the scale to amortize these investments. Ultimately the question to ask is, does a 100% DIY approach provide any differentiation, or is it a distraction?
- **Software from different vendors:** In this model, the user purchases commodity hardware and procures software from a variety of vendors (proprietary or open source). The user would act as a system integrator and then own operations as well. The burden is less than a 100% DIY case, but still substantial. DevOps, integration, testing, interop testing, bug fixing would still be the user's responsibility along with all operational tasks. The user would have to be adept at finding issues in a multi-vendor environment where finger pointing amongst vendors could kill months.
- **External system integrator:** In this model, the user would purchase the entire solution from *one* vendor, yet retain operational management of their cloud. While less burdensome that the first two models, operating an NFV cloud would be the user's responsibility.
- **Managed services:** In this last model, the user would purchase the solution from one vendor and then ask them (or another vendor) to manage it. This is the least burdensome option from an effort point of view, but may be more expensive at scale.

Clearly, there is no right answer in terms of these options. The answer may be different for different users. Choosing the wrong model could actually add to the TCO rather than cutting it.

Starting the Journey

The NFV transformation is still in early stages, but some best practices are available from users (for example see discussion by KPN and Deutsche Telekom at the SDN World Congress in 2016) .

1. **Clearly articulate goals:** It is important to focus on the right goals and communicate them internally for maximum support from the staff: A) Focus on freeing up resources for higher value functions rather than cost cutting. If the staff feels NFV threatens their jobs, they are less likely to be supportive. B) Focus on the benefits of NFV such as reduction in cost basis, increase in services, increase in customer base and increase in customer satisfaction rather than viewing NFV as cannibalizing current business. C) Communicate long-term vision *in addition* to short-term goals. D) Explain that NFV is not an optional activity. If you don't undergo NFV transformation, you run the risk of OTT competitors significantly disrupting your business.

2. **Build skills organically:** As we discussed earlier, there just aren't enough experts that you can hire, so you need to have a strong focus on building skills internally by pooling resources that are currently on different teams. Of course, these teams need to be supplemented with external skills as needed, while recognizing that there is no magic wand a vendor can wave.

3. **Agile instead of big bang:** With 4G and prior initiatives, it was not uncommon to let the specification be fully complete and then do a complete implementation. With NFV, you will need to adopt an agile philosophy without waiting for every little detail to be ironed out. Small things need to be tried out, corrected and evolved with time.

4. **Pace your journey:** As "cool" as model driven architectures, DevOps and cloud native VNFs might be, not every user has to make an immediate jump to these new technologies. In fact, several journeys are valid such as:

	Phase#1	Phase#2	Phase#3
Journey #1	Compute virtualization only	Compute virtualization + SDN controller	NFVI + SDN controller + VIM + MANO

Journey #2	Compute virtualization + SDN controller	NFVI + SDN controller + VIM + MANO
Journey #3	NFVI + SDN controller + VIM + MANO	

This above table does not talk about VNFs; even with VNFs, users could go with cloud hosted, cloud optimized, or cloud native VNFs. Similarly, users could use DevOps practices only for the development environment first, or jump into DevOps for all stages – dev/test/staging/production. And so, there are many routes to your destination.

5. **Find use case:** It is too tempting to build out a 20-50 node NFVI/MANO POC and then wait for VNF workloads. But this approach does not work – it results in orphan clouds. What does work is to find a real production workload – ideally non-mission-critical – and then create an NFVI/MANO infrastructure around it. It is perfectly fine if this is a small use case to begin with, you need some real time gains today! To use a non-telco example, in the case of Volkswagen, they picked a car configurator application for their customers in Spain and built an OpenStack cloud around it.

6. **Find executive sponsor:** It is important for a key transformation such as NFV to have top-down support. Without an executive sponsor, an initiative such as NFV is likely to wither away. Executives such as André Beijen, head of network innovation for KPN, and Deutsche Telekom Vice President Axel Clauberg have been vocal about their implementation approach to NFV.

7. **Use dedicated teams:** Rather than add another task to existing teams' burden, starting with a new dedicated team seems to be more successful. The new team can also embrace a new culture, processes and tools, as opposed to reusing existing ones. KPN, for example, assembled a new team for virtualization projects, recruiting key talent by asking for volunteers rather than assigning people.

8. **Spread the knowledge:** The initial team needs to act like a catalyst that transfers its newly acquired skills, knowledge, thinking, culture and processes to the rest of the organization. Going back to the Volkswagen example, they actively organize ongoing workshops and technology days to disseminate the knowledge across the company.

In addition, the above best practices, OPNFV is an open source community that can help you with your NFV journey. In the next chapter, we will look at what OPNFV is.

3

IT'S A BIRD, IT'S A PLANE, IT'S OPNFV!

Just as onlookers were mystified at the sight of Superman streaking through the sky, those trying to understand OPNFV through the lens of traditional open source projects might be confused by its somewhat non-traditional nature. Let's look at the driving factor for creating yet another open source project before exploring what OPNFV is.

OPNFV Driving Factors

The underlying assumption of OPNFV is that open source software and collaboration techniques are an attractive way to create parts of – if the not the entire – software stack defined by ETSI (see Chapter 1). As there is no shortage of open source networking projects to work with, this should be straightforward, correct? Actually, no! When applying these open source projects to a specific use case such as NFV, four interesting challenges arise:

- **Influencing projects**: Large open source projects have two idiosyncrasies when it comes to roadmap definition. First, there is no central product management or equivalent entity driving specific features. This decision-making happens in a much more distributed fashion. Second, these projects often target multiple use cases. So, it is a challenge to steer open source projects toward one specific use case such as NFV.
- **Composing the stack**: NFV requires software from multiple open source projects. No one project is going to own the responsibility of making sure that all these pieces of software are integrated and interoperate. Composing different flavors of stacks based on these upstream projects is especially difficult, as they approach problems from three

distinct points of view: web-scale, network, and IT. The challenge is to blend the three points of view and take the best practices from all three.
- **Testing the stack**: Finally, the stack composed from multiple projects has to be tested. While open source projects do some level of testing, it is generally limited to the scope of their own project, and no one project is going to take ownership of testing the newly composed stack – nor is any one community going to test for NFV-centric attributes.
- **Getting end user feedback:** Unless an open source project targets a particular use case, it is difficult to get in-depth user participation, which is so critical for any project to succeed.

These are some of the reasons OPNFV was created.

What is OPNFV?

OPNFV stands for Open Platform for Networks Functions Virtualization. The project was formed in September 2014, and is housed under the auspices of the Linux Foundation. Initially OPNFV's main goal was to develop an integrated and tested open source platform on which to build NFV functionality. The current mission according to the OPNFV website is:

> Open Platform for NFV (OPNFV) facilitates the development and evolution of NFV components across various open source ecosystems. Through system level integration, deployment and testing, OPNFV creates a reference NFV platform to accelerate the transformation of enterprise and service provider networks.

OPNFV builds the end-to-end stack to support NFV with verified capabilities and characteristics, establishes agile reference methodologies (requirements, documentation and propagation; continuous integration, testing, and continuous delivery), and offers a process and supporting tools for testing and validating NFVI and MANO products and solutions.

OPNFV responds to the four driving factors *directly* by organizing its project around the following three pillars, and by consciously including end users as key contributors.

Understanding OPNFV

OPNFV Project Pillars

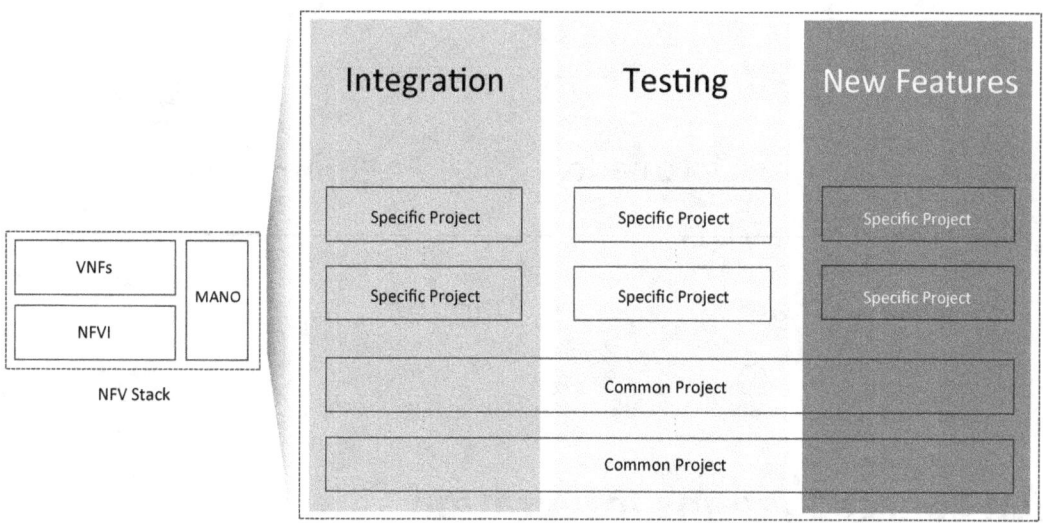

- **Integration:** OPNFV integrates a variety of open source projects to address specific NFV requirements.
- **Testing:** OPNFV tests the entire stack across a variety of NFV-specific parameters.
- **New features:** For each upstream open source project, OPNFV serves as the vehicle for NFV requirements. By actively working upstream and providing a single voice for NFV requirements, OPNFV steers these open source projects to serve the needs of NFV.

The technical work of OPNFV is organized into projects. There are projects common across all three pillars, such as CI/CD, security and documentation, and there are those specific to a given pillar.

At the time of writing of this book (April 2017), OPNFV has 51 member companies, 45+ projects, 340+ contributors, 16 test labs and 20,000+ commits (source: OPNFV community dashboard). To ensure adequate end user participation, OPNFV has 30+ members in its end user advisory group. OPNFV has completed three plugfests and organized two major worldwide summits to date. On average, OPNFV provides six-month major releases, each named after a river. Highlights from the most recent Danube release (number 4), include the introduction of stress tests, integration of the MANO stack, additional support for data plane acceleration technologies and significant improvements in the CI and test framework.

Understanding OPNFV

OPNFV Release Timeline

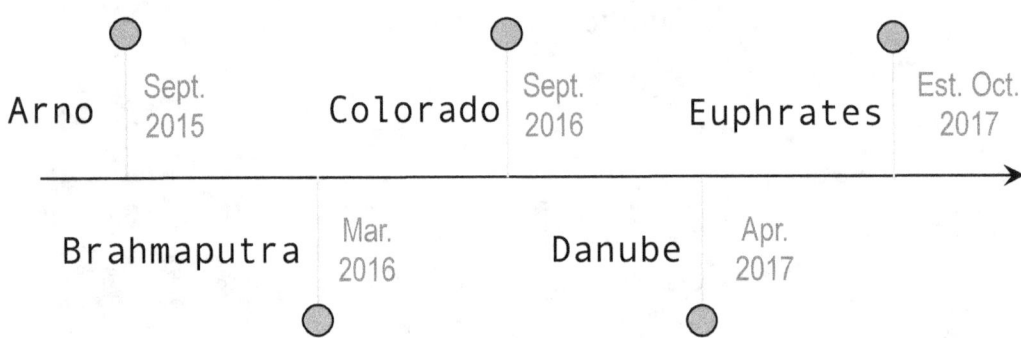

Open Source Systems Integration

Most people assume that an open source project is a group of programmers volunteering their time to develop a piece of code. However, open source projects don't need to focus on creating a piece of software. OPNFV is an integration project whose work involves testing, CI/ CD, documentation, and more.

> **OPNFV is Systems Integration as an Open Community Effort**

What OPNFV is Not

As described above, OPNFV integrates a variety of upstream open source projects, tests them, and then influences their roadmaps. However, OPNFV does not fork in order to enhance software projects or create new software functionality. That task is left to the various upstream projects from which OPNFV draws. That means that when OPNFV does enhance projects, it is done collaboratively in conjunction with the upstream community. Furthermore, even though OPNFV does extensive testing, it is not attempting to build a product. In other words, it does not prescribe any particular software combination for a particular application. That task is left to users and vendors. For these reasons, you don't necessarily download, install and run OPNFV artifacts like a traditional open source project. If you are a vendor, for the same reason, you are unlikely to have an "OPNFV distribution;" though you might have a commercial distribution of a software component that has been integrated or tested by OPNFV, or even an integrated product with several components from one of OPNFV's stacks.

OPNFV is a "Midstream" Project

Software development draws on analogies from rivers (which is very interesting given that OPNFV projects are also named after rivers) by using terms such as "upstream", "downstream", and so on. Upstream describes the code base from which a downstream code base is forked. In the context of open source, upstream describes the master repository that all contributors work on. Downstream refers to forks made to create releases and to apply specific features or bug fixes.

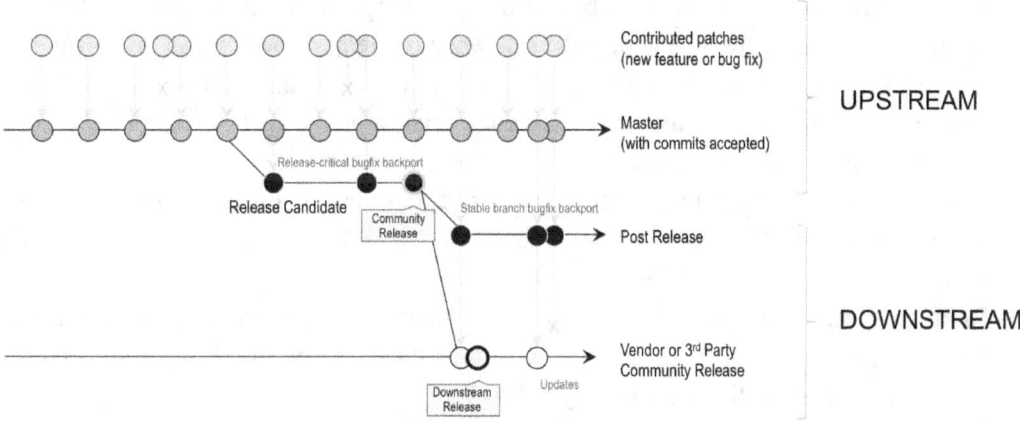

Upstream and Downstream Open Source

OPNFV is a unique project because it works both upstream and downstream. The integration and testing pillars are a downstream activity, while new feature development, on the other hand, is an upstream activity. For this reason, OPNFV may also be called a "midstream" project.

Benefits of Open Source

Since OPNFV is an open source project, a brief refresher on the benefits of open source might be in order.

Open Source Supplements Open Standards

The vehicle to gain interoperability, until recently, had been open standards. Open standards were a way for end users to, increase the set of possible vendors to avoid vendor lock-in and foster innovation. However, as Sterling Perrin makes the case eloquently in a Heavy Reading 2016 report, standards are unable to keep up with the pace of current innovation. Standards also often fail to ensure true interoperability. And since they work for long periods of time without getting customer feedback, standards sometimes end up in a wasted effort. We are by no means suggesting that standards are going away; what we are saying is that some types of standards can be replaced by open source software. In that sense, open source software supplements open standards. Case-in-point, OPNFV collaborates with standard bodies such as ETSI NFV ISG, IETF, MEF, TM Forum, and so on.

While open source, per se, is not a business benefit, the following is a list of the primary end user benefits of open source:

- **Interoperability:** Open source assures stronger interoperability than open standards because not only are the APIs open, the reference implementation is also open – and frequently created jointly by multiple companies, including competitors.
- **Innovation velocity:** The rate of innovation in open source projects typically exceeds proprietary products and standards.
- **Faster troubleshooting**: In a multi-vendor system, troubleshooting becomes a complex multi-party effort. With open source, all parties have access to the source and the ability to fix bugs, including the end user.
- **Roadmap influence:** The ability of a user to influence an open source project is higher than a proprietary product. In fact, if all else fails, the user can simply start contributing to the project and get the features they want.
- **Reduced cost:** Open source projects are not free. However, open source enables sharing R&D burden across multiple companies in the ecosystem and drives greater competition that in turn results in cost savings.
- **Transparency:** In the current age of heightened security concerns, open source transparency provides protection against malevolent features.

Multiple Approaches to NFV

OPNFV can be used collaboratively with:

- **CORD:** Central Office Re-architected as a Datacenter (CORD) is also an open source project hosted by the Linux Foundation. This project is being integrated in OPNFV as part of a new multi-edge project. CORD defines entire stacks that consist of hardware (industry standard servers, white box switches and access connectivity), along with ONOS SDN controller, OpenStack VIM and XOS VNFM. The focus is on development, as opposed to integration and testing, which makes collaboration with OPNFV very appealing. CORD has three flavors (Enterprise, Residential and Mobile)

that use the same core technology but vary VNFs and access technologies.
- **Public clouds:** Using public clouds for NFV use cases is generally impractical. However, for a subset of use cases such as vIMS, a public cloud may be an option. Also, parts of the NFV stack such as MANO software could be hosted in public clouds.

The following could be alternatives to OPNFV:
- **Proprietary virtualization solutions:** Technologies such as vSphere ESXi, proprietary MANO or SDN controllers are potential alternatives to the OPNFV stack.

So Why Should I Care Again?

As we saw in Chapter 2, NFV transformation is more about process, organization structure, and skill set acquisition than about technology. OPNFV is an unparalleled community with deep exposure to these aspects *in addition to technology*. As the remaining part of the book will show, Model driven architecture, DevOps and cloud native software are baked into the DNA of OPNFV. Participation in OPNFV could save you multiple years of false starts, and tens if not hundreds of millions of dollars in wasted expenditure.

Form a purely technical point of view, you may still be thinking that if OPNFV is not a standard open source software project with a ready-to-use single distribution that you can use, why should you spend time on it? Well, if you are looking to deploy NFV across telco or enterprise networks using open source projects, consideration of OPNFV is critical. OPNFV, in a nutshell:

- **Promotes integration and interoperability:** Open source projects are unlikely to work together out-of-the-box. OPNFV ensures combinations of different projects work together. The integration work is not trivial. Even if a user or vendor chooses not to use OPNFV artifacts, the integration work will cut down internal R&D work significantly; and those resources can be redirected to higher value activities.
- **Drives requirements:** OPNFV drives key NFV-centric features and code into relevant upstream projects.

In short, OPNFV is like a take-and-bake pizza. It's not completely cooked and ready to eat, but it has taken care of numerous labor intensive steps that reduce your time to eating, or in this case, time to NFV in production! See Chapter 10 for a deeper discussion on the benefits of OPNFV.

{Create-Compose-Deploy-Test}.Iterate

To summarize, the goal of OPNFV is to create new features, integrate a variety of upstream open source projects to compose NVF stack(s), deploy and test them. This entire cycle gets repeated over and over again, triggered by any upstream component change.

In subsequent chapters, we will look at each of these four stages (create, compose, deploy, test) in more detail.

4

UPSTREAM PROJECTS IN OPNFV

As described in the previous chapter, OPNFV is a "midstream" project that draws from a number of upstream open source projects. OPNFV integrates and tests combinations of different projects, and also influences these projects to align them with NFV requirements. When the community does contribute code, it's usually contributed directly to the relevant upstream project. This chapter provides an overview of the main upstream projects used in OPNFV.

Upstream Projects Used in OPNFV

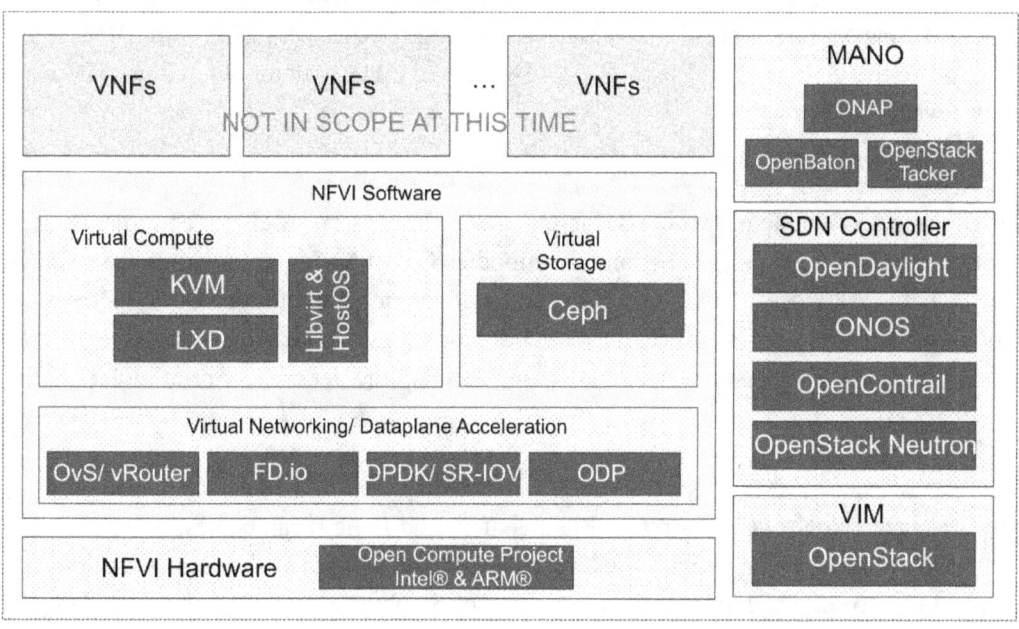

As you see from the above diagram, in several cases, OPNFV integrates multiple software projects for the same purpose. In this way, OPNFV enables choices for end users through multiple options, and it is not OPNFV's role to recommend one technology over another.

A common temptation when comparing open source projects is to focus on features and functionality. An equally, if not more important aspect to look at, however, is the community, and its constitution, structure, governance, momentum, funding and development methodologies. You will see wide variations in these factors among the projects discussed below.

NFV Management and Orchestration (MANO)

The following MANO projects are currently integrated with OPNFV: OPEN-O (part of ONAP), OpenBaton and Tacker.

ONAP

The Open Network Automation Platform (ONAP) is a Linux Foundation end-to-end service orchestration project. It is the result of a merger of two projects: OPEN-O and Open Source ECOMP (open source version of AT&T's ECOMP project). AT&T and China Mobile are driving ONAP with a diverse group of members. Platinum members include Amdocs, AT&T, Bell Canada, China Mobile, China Telecom, Cisco, Ericsson, Gigaspaces, Huawei, IBM, Intel, Reliance Jio, Nokia, Orange, TechMahindra, VMware, and ZTE, with more to come. There are numerous silver members as well.

The ONAP project is unique in two ways. First, it started as a pair of user-driven projects, as opposed to a vendor-driven project. Second, it embodies Cloud Native architecture from the get-go by using a microservices architecture and an agile development methodology. As a testament to agility, ECOMP, with over 8 million lines of code, took only 1.5 years to develop with just 300 developers! Because a combined release is not yet available, let's take a brief look at both projects separately.

ECOMP (Enhanced Control Orchestration Management and Policy) can be best described as a MANO++ project. ECOMP automates the design and delivery of network services that run on a cloud. In addition to service delivery and automation of SDN tasks, ECOMP also automates many service assurance, performance management, and fault management tasks.

ECOMP

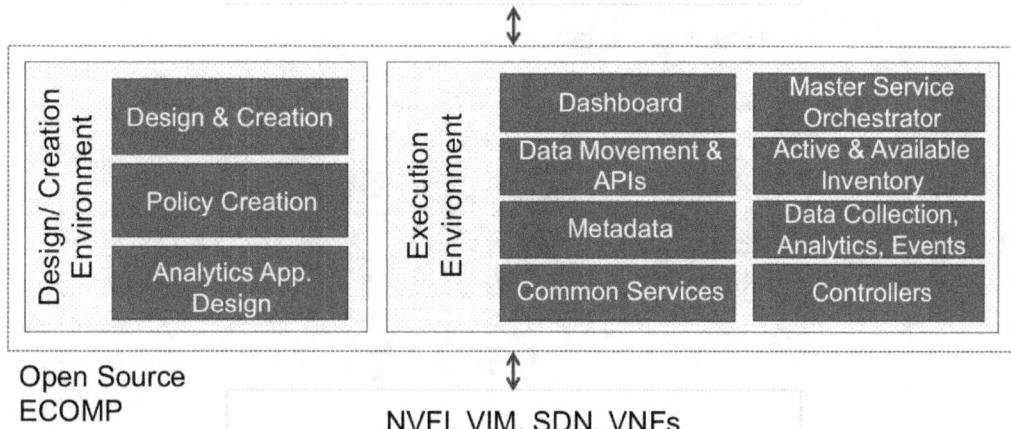

ECOMP adds several capabilities over and above the ETSI MANO architecture. It addresses the entire service delivery lifecycle using model driven architectures. For example, it has a rich service design environment for service creators to build services using a collaborative, catalog-driven self-service design studio. The design studio also creates service, VNF, and infrastructure descriptions that are richer in scope than what's required by the ETSI architecture by adding numerous pieces of metadata. NFVO and VNFM functionalities are enhanced for greater control over NFVI, VIM, and SDN controllers, and for faster on-boarding of new VNF types. ECOMP supports YANG, TOSCA, OpenStack Heat and other modeling languages. Finally, the project includes FCAPS (Fault, Configuration, Accounting, Performance, Security) functionality to provide greater control over closed loop automation (in the ETSI architecture this functionality resides in the EMS – Element Management System). While OpenStack is the primary VIM (Virtual Infrastructure Manager) supported, Open Source ECOMP can be extended to other VIMs too.

OPEN-O is a Linux Foundation open orchestration project that combines NFV MANO and connectivity services orchestration over both SDN and legacy networks. OPEN-O employs a model-driven automation approach, and has adopted standard modeling languages YANG (for networking devices) and TOSCA (for services).

The OPEN-O architecture is partitioned into three main orchestration functions: global services orchestrator (GSO), SDN orchestrator (SDN-O), and NFV orchestrator (NFV-O), along with a set of common services. OPEN-O is built upon a microservices architecture for extensibility, and features a modular design that supports multiple VIMs, VNFMs, SDN Controllers, and legacy network and element management systems.

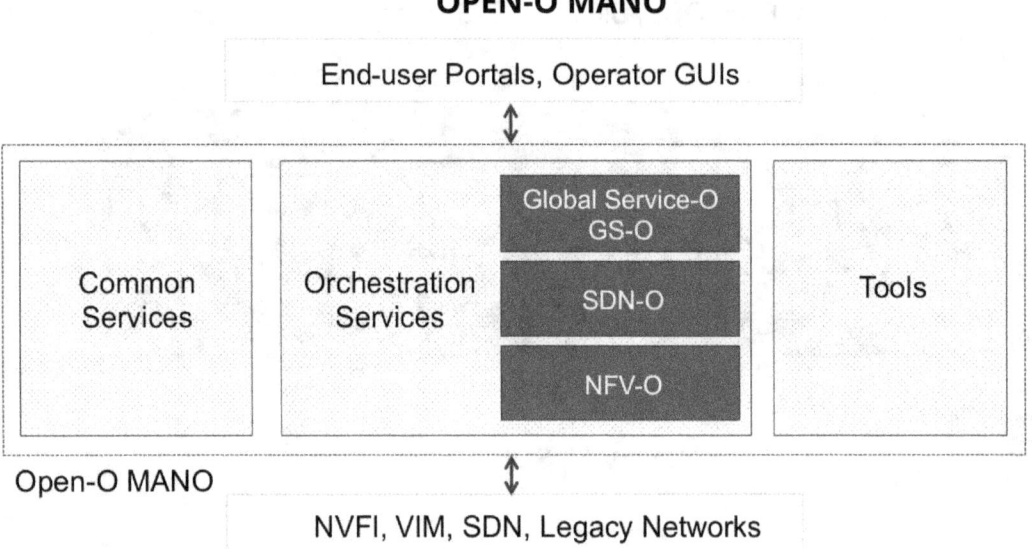

In the Opera project, OPNFV integrates the OPEN-O project, with Juju and Tacker utilized as VNFMs.

OpenBaton

A MANO project started by Fraunhofer Fokus, the largest research group in Germany, OpenBaton was designed with modularity, extensibility and interoperability in mind.

Now on its 3rd release, OpenBaton enables you to use just a couple of modules: NFVO and a messaging bus module – or the entire set of components, such as the fault management system, autoscaling engine, generic VNF manager (VNFM) or a Juju VNFM. It also includes support for multiple VIMs, with OpenStack being the primary one; Docker support has been demonstrated as well. OpenBaton supports network slicing, which will become important with 5G, and has a VNF marketplace.

OPNFV has integrated OpenBaton into the platform under the Orchestra project.

OpenBaton MANO

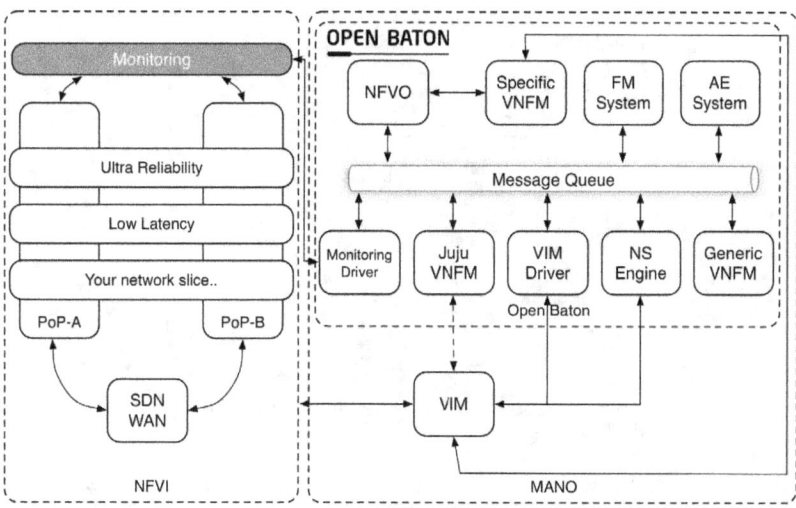

OpenStack Tacker

Tacker is an official OpenStack project that started off as a generic VNFM but has since expanded its scope to provide an NFVO as well, making it a full-fledged MANO project.

Being a part of OpenStack, it is also well integrated with other OpenStack components such as Nova, Horizon, Heat, Keystone, and so on. The main contributors are Brocade, Infinite, NEC, 99cloud, Huawei, Imagea, Nokia, and China Mobile. Because Tacker can be used as just a VNFM, it can also coexist with other open end-to-end service orchestration projects such as OPEN-O and ONAP.

Tacker MANO

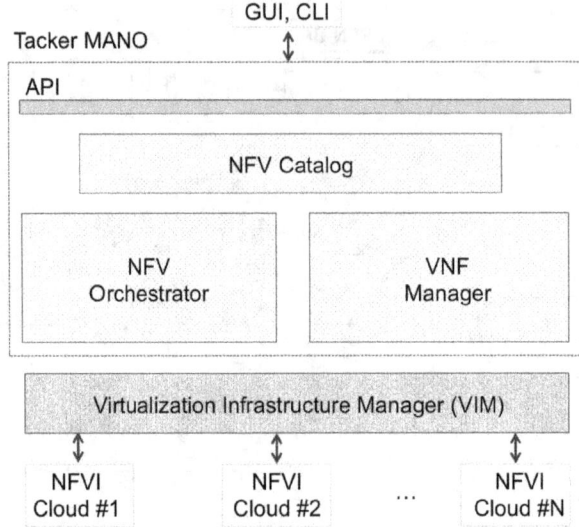

In addition to these three MANO projects, Open Source MANO (OSM), originally from Telefónica and now housed under ETSI, could be integrated into OPNFV at some point.

Virtualized Infrastructure Manager (VIM)

The VIM is responsible for creating and managing infrastructure resources such as virtual compute and virtual storage. The VIM can work with an SDN controller (see next section) to create virtual networks. (Strictly speaking, SDN controllers are part of the VIM, but we are breaking out each of these components for the sake of clarity.)

OpenStack

Currently OPNFV uses OpenStack as its primary VIM. In the context of NFV:

> OpenStack® is open source software for managing telecommunications infrastructure for NFV, 5G, IoT and business applications. Global telecoms including AT&T, China Mobile, Orange, NTT DOCOMO and Verizon deploy OpenStack as an integration engine with APIs to orchestrate bare metal, virtual machine and container resources on a single network. OpenStack is a global community of more than 70,000 individuals across 183 countries supported by the OpenStack Foundation. Visit the Telecom and NFV page to learn more. (Source:

OpenStack Foundation).

OpenStack, one of the largest open source projects in the world, is a seven-year-old project and its 15th release, Ocata, was developed by 1,925 community members from 285 organizations. Sponsors include Platinum members such as AT&T, Canonical, Huawei, IBM, Intel, Rackspace, Red Hat and SUSE, as well as a large number of Gold and other sponsors. In a recent Heavy Reading survey, 86% of global telecoms stated that OpenStack is either important or essential to their success, and over 60% are also using or testing OpenStack for NFV. Perhaps more interesting, over 21% of those surveyed, plan to get OpenStack as part of OPNFV!

OpenStack itself is an umbrella project, with numerous projects underneath. A detailed description of each project is outside the scope of this book, but the core services are as follows. Interestingly, there are some OpenStack projects that are relevant to NFV, but not integrated into OPNFV at this time; the last column shows if the project has been integrated or not.

OpenStack Core Projects

Service	Project	Integrated in OPNFV?
Compute	Nova	Yes
Block Storage	Cinder	Yes
Networking (full SDN controller *or* just an API layer)	Neutron	Yes
Identity	Keystone	Yes
Image Service	Glance	Yes
Object Storage (API layer *and* software-defined storage)	Swift	Optional

OpenStack includes numerous optional projects, and several of them are relevant to NFV.

The OpenStack Project Navigator lists each project and provides tags for age, maturity and adoption. Age indicates how long the project has been in development, adoption reflects the percentage of production deployments as reflected via bi-annual user surveys. There are several additional maturity indicators, including rolling upgrade support, vulnerability management team support, and documentation. These tags are provided for your project evaluation. Typically, core projects are more mature than optional services, so as a user, you have to choose carefully.

OpenStack Optional Services Relevant to NFV

Service	Project	Integrated in OPNFV?
Alarming service (alarms and notifications based on metrics)	Aodh	Yes
Orchestration	Heat	Yes
Governance	Congress	Yes
VNFM/ MANO	Tacker	Yes
Root cause analysis (linked to OPNFV Doctor)	Vitrage	Yes
Reservation-as-a-service (linked to OPNFV Promise)	Blazar	Yes
Networking automation across Neutron in multi-region deployments	Tricircle	Yes
Clustering service	Senlin	Yes
Dashboard	Horizon	Yes
Telemetry	Ceilometer (used by OPNFV Barometer project)	Optional
Workflow service	Mistral	Optional
Monitoring-as-a-service	Monasca	Optional

Time series database as a service	Gnocchi	No
DNS service	Designate	No
Distributed backup restore and disaster recovery as a service	Freezer	No
Data protection as a service	Karbor	No
Distributed SDN controller (can replace Neutron SDN controller)	Dragonflow	No

Beyond optional services are community projects, affiliated with OpenStack. Some of them are relevant to NFV.

OpenStack Community Projects Relevant to NFV

Service	Project	Integrated in OPNFV?
Model-driven, extensible framework for NFV networking service (can replace Neutron API layer) linked to OPNFV NetReady	Gluon	Yes
Centralized service for multi-region deployments linked to OPNV Multisite	Kingbird	Yes
Parser for TOSCA simple profile in YAML	TOSCA-Parser	Yes
Integrated network service orchestration	Astara	No
Group based policy	GBP	No

The latest OpenStack release, Ocata, has important new features for NFV, and we'll cover more about these topics as we progress through the book. The key takeaway is that OpenStack views NFV as an important use case, and the OpenStack Ocata release has important new features, such as improvements to Nova, Vitrage, Congress, Neutron, TOSCA-Parser, Heat-Translator, and the introduction of DragonFlow and Tricircle into OpenStack as optional services.

VIM Options Beyond OpenStack

In addition to OpenStack, the Danube release has experimental support for a VIM based on Kubernetes, an open source container orchestration project. In the words of the official site:

> Kubernetes is an open source system for automating deployment, scaling, and management of containerized applications. It groups containers that make up an application into logical units for easy management and discovery. Kubernetes builds upon 15 years of experience of running production workloads at Google, combined with best-of-breed ideas and practices from the community.

Kubernetes enables containerized VNFs to be used in NFV deployments. The use of Kubernetes in the Danube release does not include any SDN controller integration, nor is it integrated with any other NFV-centric project. However, we expect Kubernetes integration to deepen over subsequent releases. OPNFV OpenRetriever is one such project that aims to provide Kubernetes integration and testing for the Euphrates release.

Software Defined Networking (SDN) Controller

The SDN controller is a control plane that is responsible for programming physical and virtual networking elements. Before SDN, switches or routers were configured using routing protocols that did not allow fine grained control. SDN controllers have a northbound interface that connects with MANO and VIM components, and a southbound interface to physical and virtual networking switches and routers. While northbound interfaces have not been standardized, a variety of standards exist for the southbound interface, such as OpenFlow, OpFlex, Netconf, P4, ovsdb, and so on.

SDN controllers are responsible for setting up overlay networks. An overlay network is a virtual network built on top of a physical network that connects virtual machines. These overlay networks can go through a router or a gateway to communicate with the external world.

SDN controllers offer numerous benefits:
- The entire control plane is centralized for ease-of-management, though it can still be

distributed for HA and scaling
- Policies are centralized, so conflicts are eliminated
- Rapid deployment of network-related changes through the centralized control plane is manageable using scripting and other programmatic means
- Event-based reconfiguration of networking elements is possible

OPNFV integrates five SDN controllers:

OpenStack Neutron

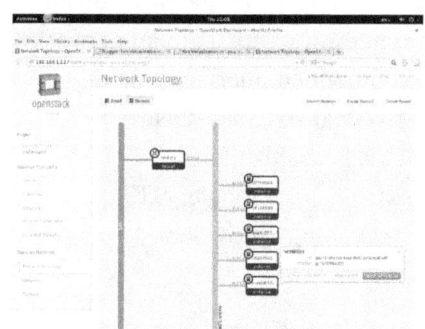

OpenStack Neutron is both an API translation layer *and* an SDN controller. If used only as an API layer, it utilizes *core plugins* to connect to third-party SDN controllers through their respective northbound interfaces. If used as both an API layer and an SDN controller, Neutron manages OVS (see *NFVI Virtual Switching* below) and external physical switches through *drivers*. Typically, Neutron is not used as an SDN controller in complex environments, and instead it is used only as an API translation layer. Also, if you weren't confused by the dual personality of Neutron, the project actually also has *service plugins* that would traditionally be thought of as VNFs, such as load balancer as a service (LBaaS), firewall as a service (FWaaS), VPN as a service (VPNaaS) and so on.

OpenDaylight

Like OPNFV, OpenDaylight (ODL), is also a Linux Foundation project. It is a full blown modular SDN controller that caters to multiple use cases such as NFV, IoT, and enterprise applications. It supports numerous southbound interfaces to manage virtual and physical switches (OpenFlow, Netconf and other protocols). For the northbound interface to OpenStack or other

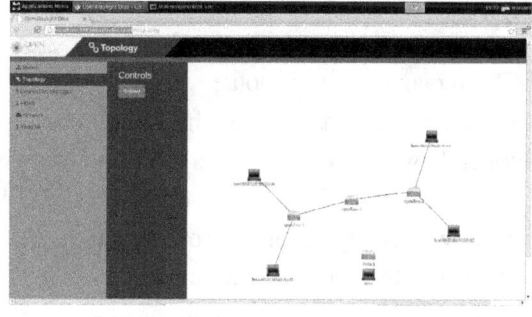

orchestration layers, ODL uses YANG (a standard modeling language) models to describe the network, various functions, and the final state. The ODL community is large, with Brocade, Cisco, Ericsson, HPE, Intel, and Red Hat being just a few of the companies supporting the initiative.

ONOS

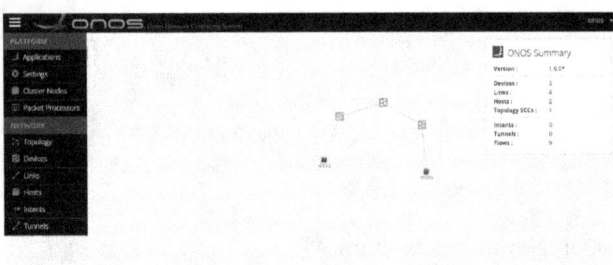

ONOS is a modular SDN controller specifically built for service providers with carrier-grade considerations such as scalability, high performance, and high availability in mind. The core of ONOS is distributed, so it can scale horizontally. Northbound interfaces are based on an intent framework with a global network view. Southbound interfaces include OpenFlow and Netconf to be able to manage both virtual and physical switches. ONOS is managed by the Open Networking Lab (ON.lab), a member-funded organization with 17 key members that include operators such as AT&T, China Unicom, Comcast, Google, NTT Communications, SK Telecom, and Verizon, and a number of technology providers.

OpenContrail

OpenContrail is an SDN controller that was open sourced by Juniper. It targets both cloud and NFV use cases for enterprises and service providers. Unlike the other two SDN controllers, OpenContrail uses BGP and Netconf as the primary southbound interfaces to manage physical switches and routers, and XMPP to manage virtual router  elements in compute nodes. Given the XMPP choice instead of OpenFlow, OpenContrail does not interoperate with OVS and installs a vRouter on each compute node. For overlays, OpenContrail leverages MPLS protocols over GRE, UDP or VXLAN. The OpenContrail architecture has three types of controllers that all scale horizontally: control node cluster, configuration cluster, and analytics cluster.

OVN

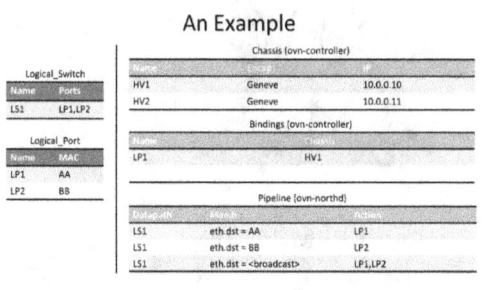

An Example

OVN stands for Open Virtual Network. It is being developed by the same team that developed the OVS. The project aims to add virtual network abstractions such as virtual L2 and L3 overlays and security groups. OVN is dissimilar from the above projects in that it focuses narrowly only on virtual networking rather than being a full blown SDN controller. So, it does not try to solve the problem of managing physical switches. In this way, the code is lighter and more focused, but solves a narrower problem.

NFVI Compute

This layer provides the datapath for running virtual machines or containers. These compute instances are scheduled, i.e. created and torn down, by OpenStack Nova.

KVM

Kernel-based virtual machine (KVM) is a hypervisor that creates virtual machines, and is part of the Linux kernel.

KVM

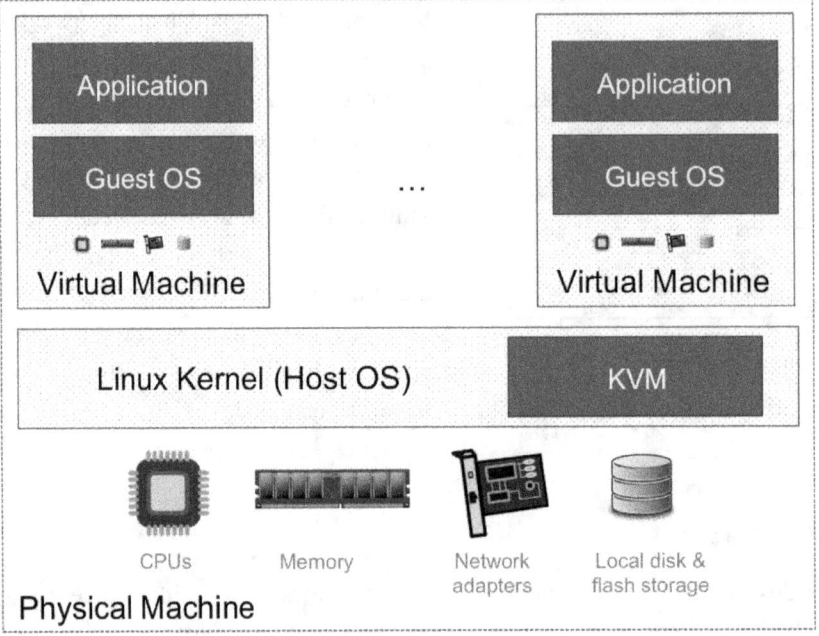

KVM provides a virtual machine that looks just like a physical machine to the operating system and associated applications – it has virtual CPUs, memory, networking ports, and SSD or HDD storage. Since each virtual machine has its own operating system instance, the isolation and security characteristics are almost as good as separate physical machines. KVM is a mature, 10-year-old project.

LXD

OPNFV also integrates LXD, a system container technology from Canonical that, at a high-level, is similar to Docker. It uses the same low-level isolation mechanisms as Docker to provide isolation of containers.

LXD

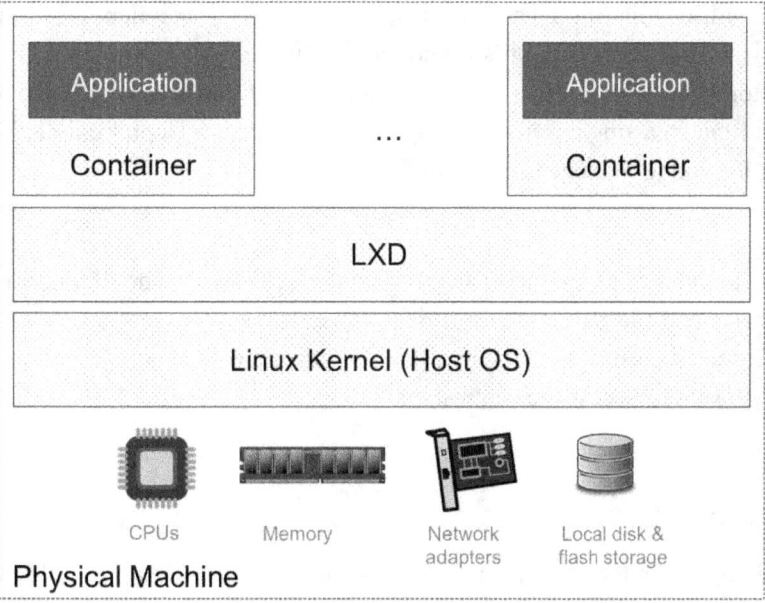

Containers virtualize the operating system as opposed to the entire machine. As a result, containers are much lighter than virtual machines, but suffer from isolation and security issues by virtue of sharing a common kernel. As an aside, for this reason many users run containers inside virtual machines. Given their small footprint, containers are popular in cloud native micro-services-based software architectures.

libvirt/ HostOS

OPNFV supports Ubuntu, CentOS and SUSE as host operating systems that run the hypervisor. libvirt is a virtualization API to manage hypervisors running on the Host OS. It communicates with the VIM on one side and KVM or LXD on the other side.

NFVI Storage

According to the OpenStack user survey, Ceph is the most popular external block storage software used in OpenStack deployments. For this reason, Ceph is also integrated with OPNFV. Ceph is software defined storage that scales horizontally by adding storage servers. The software was open sourced by Red Hat, and now has contributors from Red Hat, ZTE, Mirantis, SUSE, XSky, Digiware and Intel. According to the official website:

Ceph is a unified, distributed storage system designed for excellent performance, reliability and scalability. Ceph's foundation is the Reliable Autonomic Distributed Object Store (RADOS), which provides your applications with object, block, and file system storage in a single unified storage cluster – making Ceph flexible, highly reliable and easy for you to manage.

With cloud native architectures, the need for persistent block storage should diminish with database-as-a-service and filesystem-as-a-service; however, as VNFs are not cloud native yet, persistent block storage is useful for high availability where a pair of active-active or active-passive VNFs may want to access the same state.

NFVI Virtual Switching

These virtual switch and router elements provide the virtual networking or overlay layer to connect virtual machines or containers. These elements are managed by the SDN controllers.

OVS

The Open vSwitch (OVS) project is a Linux Foundation project. It was started in 2009 and is broadly used in the industry. It implements a virtual switch that runs in the host Linux operating system to provide virtual networking between virtual machines.

OVS is a multilayer virtual switch that can be distributed across multiple physical machines, and includes security, monitoring, QoS, VLAN, VXLAN, GRE, NetFlow, sFlow, NIC bonding and automated control features. The reference implementation can be managed via OpenFlow. OpenStack Neutron, ONOS, and ODL all support OVS.

vRouter

OpenContrail comes with its own virtual networking technology, a vRouter, that runs in the hypervisor. It is similar to OVS, except that it can perform forwarding too, hence the name vRouter. Instead of OpenFlow, OpenContrail relies on XMPP to manage the vRouter.

FD.io

A high-performance alternative to OVS, the core engine of the Fast Data project (FD.io) is a vector packet processing engine (VPP) that came from Cisco.

VPP processes a number of packets in parallel instead of one at a time. This spreads the overhead of lookups and computation across an entire set of packets contributing to the added efficiency. VPP exposes a high-performance low level API. For software stacks that need to use a higher-level mechanism to communicate with VPP, another FD.io technology called Honeycomb translates YANG models exposed via Netconf/Restconf to VPP APIs. A controller which supports Netconf/YANG, such as ODL, can "mount" the Honeycomb Management Agent to communicate with the VPP.

Independent testing has shown FD.io throughput to be about 5x better than OVS with DPDK (see next section) when forwarding to 2,000 IPv4 addresses and about 39x better when forwarding to 20,000 IPv4 addresses. FD.io, you guessed it, is also a Linux Foundation project and the Platinum members of FD.io include Cisco, Ericsson and Intel, with 12 additional Gold and Silver members.

Data Plane Acceleration

Given the importance of performance, as measured by packets per second, latency or throughput, special technologies are required to optimize data plane performance. This requirement is unique to NFV and not typically shared by enterprise use cases that are generally compute intensive. OPNFV integrates three primary data plane acceleration technologies. The OpenStack Ocata release with Nova's Placement API will enable better control over the placement of virtual machines associated with specific data plane acceleration resources.

DPDK and Other Performance/ Scaling Features

DPDK, another Linux Foundation project, is a set of libraries that bypass the kernel and provide polling mechanisms, instead of interrupt based operations, to speed up packet processing. In a recent study, OVS with DPDK showed a 75% improvement in throughput over plain OVS.

The technology is available on multiple processors and is being promoted by Intel® as part of a broader portfolio of Enhanced Platform Awareness (EPA) technologies aimed at accelerating the data plane. The main technologies in EPA other than DPDK are huge pages, NUMA pinning and SR-IOV. Huge pages improve VNF efficiency by reducing page lookups, NUMA pinning ensures that the workload uses memory local to the processor, and SR-IOV enables network traffic to bypass the hypervisor and go directly to the virtual machine.

ODP

The OpenDataplane (ODP) project, supported by the Linaro Networking group and its 13 member companies, aims to create a set of standard cross-functional APIs for networking data plane acceleration. Hardware vendors love to create acceleration features such as security, switching offload, and others. However, no user wants to use proprietary features. ODP bridges this divide by providing APIs that applications can utilize in a standard way, and that vendors can adapt to their specific hardware acceleration features.

NFVI Hardware

OPNFV tests its software against proprietary and open source hardware based on both Intel® and ARM® architecture. One notable open source hardware project is Open Compute Platform (OCP). OCP is different from traditional open source software projects since it is all about hardware! OCP was started by Facebook to bring their efficiency, cost and power innovations to a broader community. This community designs compute servers, storage servers, networking equipment and entire racks in the open source. The telco working group within OCP specifically addresses the needs of telecom operators. The group has a new carrier grade 19" rack level system design that accommodates compute, storage, networking and GPU sleds.

Out-of-Scope

At this time, functional testing of VNFs, VNF integration or VNF development is out of scope; nevertheless, some tests do take advantage of open source VNFs such as the vIMS Clearwater project. Additionally, the OPNFV Samplevnf project aims to use sample open source VNFs, that mimic the behavior of real VNFs, for benchmarking and performance optimization.

In the next chapter, we look at how OPNFV collects requirements for these various projects and influences them to get these features coded.

5
OPNFV UPSTREAM CONTRIBUTIONS

OPNFV has several projects that collect NFV-centric requirements for the various upstream open source communities and influence those projects, even contributing code to upstream communities and where needed. These efforts aim to recreate the same level of service assurance, Quality of Service (QoS) and availability on a virtual infrastructure that is currently available on a physical infrastructure.

A typical upstream contribution *project* has deliverables such as:
- **Requirements gathering:** Use cases, architecture, requirements analysis, and so on
- **Documentation:** Release notes, installation guides, user guides, platform overview, configuration guides, and so on
- **Upstream contributions:** Working with upstream communities to define blueprints and contribute code that fulfils the requirements gathered by the project

Below is a list of all upstream contribution projects (as of the OPNFV Danube release) and a brief description of each. More details on OPNFV projects are available on the wiki. These projects fall into the following general categories:
- Service assurance and availability
- Easing integration of upstream projects
- NFVI/VIM/VNF deployment and lifecycle management
- Documentation and security

Project Details

Projects related to **service assurance and availability** are:

Project	Description
Availability (HA for OPNFV)	This project creates APIs and requirements for high availability (HA) in carrier-grade NFV scenarios. The project addresses HA at three different layers: hardware HA, virtual infrastructure HA and service HA. As an example of the project's success, during the Colorado release of OPNFV, this project identified a number of gaps that were successfully closed in OpenStack's Mitaka release, while some others were deferred to the Newton release.
Barometer (Software fastpath service quality metrics)	There is no shortage of open source data collection monitoring tools. Barometer takes an NFV-centric approach to this problem by capturing statistics and events from the NFVI layer in order to detect faults and enforce service level agreements (SLAs). Barometer passes this data on to higher level fault management systems. The project uses `collectd` for this purpose, and has an extensive set of plugins. These plugins range all the way from IPMI, BIOS, OVS, and DPDK to platform monitoring. Information gathered can be reported to higher level tools using a variety of interfaces, including standard telco interfaces such as SNMP.
Doctor (Fault management)	Doctor creates fault management and maintenance systems to ensure high availability for VNFs (see the next section) to match what operators are used to with physical network functions.
VES (VNF Event Stream)	While Barometer focuses on VNFI, VES gathers VNF event streams using a common model and format to enable the management of VNF health and life cycle. Currently end users have to deal with events and statistics from VNFs in a variety of different formats using different standards (for example, SNMP vs. 3GPP or CSV vs. XML) in order to implement service assurance. This increases engineering costs and delays the introduction of new VNFs. VES aims to streamline this by creating three components: an agent for mapping source telemetry to the VES event format, collector for receiving and displaying events, and a VES data schema to support these two activities.

Understanding OPNFV

Projects aimed at **easing the integration of various upstream projects** into OPNFV include:

Project	Description
Forwarding Graph (OpenStack based VNF forwarding graph)	This project plans to work with OpenStack networking-sfc and Tacker, along with ONF OpenFlow to demonstrate that this combination of technologies can be used to dynamically set up VNF forwarding graphs (VNFFG), aka service function chaining. It further demonstrates that traffic from different tenants can flow through distinct service chains (that is, a sequence of VNFs).
Movie (Model oriented virtualized interface)	OpenStack has an IaaS-centric API. However, that burdens the MANO layer with a lot of detailed work. The Movie project is creating a more abstract VIM northbound API (NBI) as an alternative to the existing OpenStack API, which could simplify MANO's job in areas such as resource access, connection generation, flow identification, policy operation and so on.
NetReady (Network readiness)	Given that NFV is first and foremost a network-centric workload, the NetReady project performs the important role of understanding the gaps in current OpenStack networking models and APIs, as they pertain to carrier grade needs. Requirements such as L3 only, WAN connectivity, and support for legacy networks, and so on for specific NFV use cases means that the current OpenStack networking architecture needs to be evolved. In addition to OpenStack Neutron, NetReady works with the new OpenStack Gluon (Model-Driven, Extensible Framework for NFV Networking Service) project.
NFV-KVM	The NFV-KVM project focuses on the KVM hypervisor in the NFVI and develops requirements and collaborates with the upstream community to achieve this integration. By using real-time KVM, the community has shown a 10x improvement in small packet performance.
ONOSFW (ONOS framework)	The ONOSFW project develops requirements for the ONOS SDN controller in OPNFV.

OpenContrail virtual networking for OPNFV	This project develops requirements for the OpenContrail SDN controller in OPNFV. [Author's note: OPNFV generates requirements for the OpenDaylight (ODL) SDN controller as well, it's just that there isn't a dedicated project in OPNFV.]
Open vSwitch for NFV	This project develops requirements, collaborates with the upstream community, and creates functional and benchmarking tests to help integrate OVS in OPNFV. One main area of focus has been performance, and moving OVS from the kernel to the user space and taking advantage of data plane acceleration has resulted in 10x improvement in packet throughput for small packet sizes.
Opera (OPEN-O integration)	The Opera project develops requirements for OPEN-O MANO support in OPNFV with an emphasis on APIs and data models. The VNFM used is Juju.
Orchestra (OpenBaton integration)	The Orchestra project develops requirements for OpenBaton MANO support in OPNFV with an emphasis on APIs and data models. The VNFM used is Juju.
SDN VPN (SDN distributed routing and VPN)	This project develops requirements for OpenStack BGPVPN in OPNFV. This project enables the integration of layer-3 networking services with wide-area-networks (WAN).
SFC (Service function chaining)	The SFC project develops requirements, documentation and infrastructure to integrate the upstream ODL SFC implementation project in OPNFV.

Projects around **NFVI/VIM/VNF deployment, lifecycle management,** and related topics:

Project	Description
Copper (NFVI deployment policies)	A user's intent for a service gets orchestrated to VNF deployments, which then get orchestrated to virtual infrastructure. It is possible that a user's intent in terms of affinity for a geography, partitioning, security or specific HA requirement may get lost in translation by

	the time the infrastructure gets provisioned. Copper ensures against this problem by working with two upstream projects: OpenStack Congress (policy-as-a-service) and OpenDaylight Group-Based Policy (GBP). The eventual goal is to ensure policies are part of the VNF package or metadata and to ensure these policies are enforced at every layer across diverse infrastructure managers.
Domino (Template distribution service)	Centralized orchestration of VNFs and the underlying infrastructure may not always be possible. For example, carrier networks may span geographies, or operators may undergo mergers and acquisitions resulting in heterogeneous orchestration tools. These situations require a top-down layer that can take a template describing service models and policies, and partition that template into specific templates for each local orchestration and controller tool. This is the role of the Domino project. Domino converts policies to TOSCA and distributes respective templates using a pub/sub system while taking dependencies into account. The scope of the project includes defining functionality, APIs, test/integration and debugging/tracing.
ENFV (Edge NFV)	In vCPE, vPE (provider edge), IoT, or other use cases, it makes sense to split the NFVI between a centralized cloud and edge compute nodes. This is also called fog computing. Splitting NFVI in this manner creates unique challenges. The goal of this project is to specify requirements for edge NFV in areas such as tunneling, the ability to handle many small "data centers", remote node monitoring, data plane optimization, and so on. The project collaborates with other OPNFV projects and influences upstream OpenStack, ODL, and OVS projects.
Escalator (Smooth upgrade)	With NFV, upgrading the NFVI and VIM becomes more complicated than fixed function boxes. The Escalator project plans to define requirements for smooth upgrades for major or minor version changes in a manner that does not unreasonably degrade VNF uptime. The project identifies parameters such as the maximum duration of upgrade or rollback, maximum duration of VNF interruption, and mechanisms to prepare an upgrade. More recently, the project has also created test cases to test upgrades.

Models (Model-driven NFV)	Different VNFMs use different techniques and data models. The goal of the Models project is to disseminate and promote convergence of information and/or data models pertaining to VNFM. Activities range from creating use case tests to comparing VNF packages for different VNFMs and fleshing out end-to-end lifecycles for services and VNFs. Finally, the project also acts as a liaison between OPNFV and other NFV standard bodies such as BBF, ETSI, MEF, OASIS, ONF, TM Forum, and so on.
Multisite (Multi-site virtualized infrastructure)	NFV use cases require that services span infrastructures that are distributed across geographical locations, more so than enterprise use cases. This creates new requirements for core OpenStack services such as Nova, Cinder, Neutron, Glance, Ceilometer and Keystone in areas of network management across sites, multi-site image replication, global and per-site quota management, and so on. This project documents these requirements and influences various upstream projects. Multisite integrates with OpenStack Kingbird (Centralized service for multi-region deployments) and OpenStack Tricircle (Networking automation across Neutron in multi-region deployments). In the OpenStack Ocata release, cells v2 in Nova makes it easier to scale and have multiple compute environments managed together. Multisite is also the magic that enables, for example, a 99.999% available service on infrastructure that is only 99.9% available. By having a service straddle availability zones or even geographic regions, a service can meet carrier grade requirements.
Promise (Resource management)	There are situations when the MANO layer might want to reserve compute, storage and network resources for a certain duration of time. Before this project, OpenStack had no such mechanism. The OPNFV Promise project solved this gap by helping create the Blazar (reservation-as-a-service) project in OpenStack.

Projects around **documentation and security**:

Project	Description
Documentation	The main task of this project is to create the required documentation for each release. Additionally, the project also develops guidelines and tooling for documentation across all OPNFV projects and maintains documentation libraries.
Moon (Security management)	The Moon project is in the process of working with the upstream OpenStack Keystone and Congress projects to improve the isolation, protection and interaction between VNFs. The project does so by identifying gaps in OpenStack and ODL and contributing features to upstream projects around authorization, logging, network enforcement, storage enforcement, and so on. Moon also enables OPNFV Copper to ensure that security policies are being met. OPNFV has also received the Core Infrastructure Initiative Best Practices Badge. Core Infrastructure Initiative (CII) is a Linux Foundation project to fund and support critical open source projects. Their goal is to fortify key projects to prevent security vulnerabilities such as Heartbleed. CII issues Best Practices Badge to showcase an open source project's commitment to security.

The above projects are all executed by small teams that are generally distributed around the world. Each project has a project technical lead (PTL), committers, and contributors. Teams work year-round, communicating via collaboration tools (described in Chapter 6), weekly meetings, and IRC. To engender periods of intensive collaboration, there is a design summit once a year concurrent with the OPNFV summit. Each year there are also multiple hackfests, at which team members can collaborate face-to-face. These are usually concurrent with plugfests (see Chapter 8).

Project Analytics

To learn how these projects are doing, a variety of statistics about each project are available on the OPNFV analytics site.

OPNFV Project Analytics

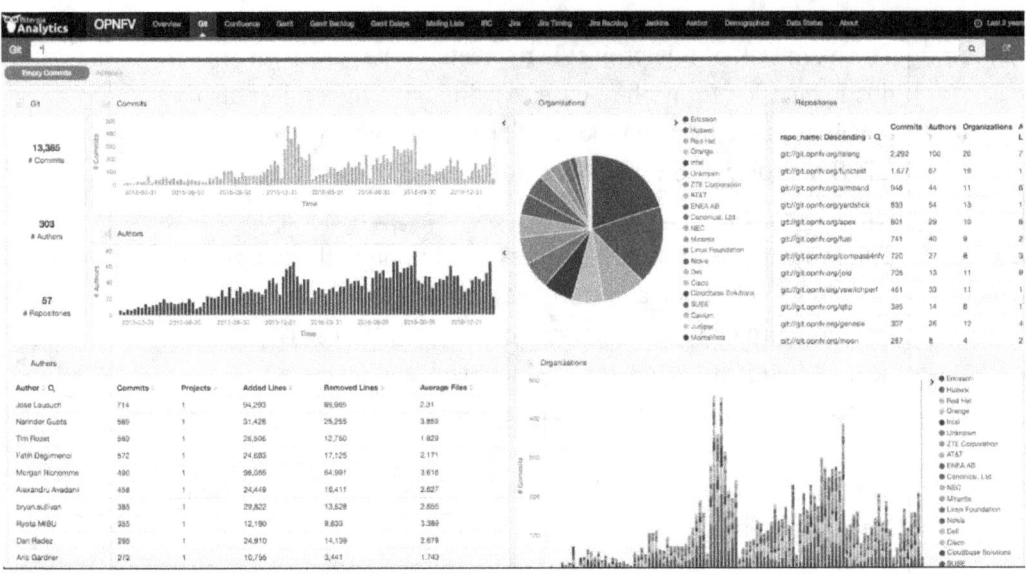

Success Story: OPNFV Doctor with OpenStack Vitrage

An example of a successful OPNFV project creating carrier-grade features in upstream projects is Doctor. The project was proposed by NTT DOCOMO in November 2014 to ensure that fault management in the virtualized world would be just as good as in the physical world. At the time of the project proposal, fault management in physical networks was very well understood and efficiently implemented. A failure would get detected by the monitoring system, and automatic failover would take place. However, the combination of virtualization and cloud-hosted VNFs broke this mechanism.

Given the state of OpenStack at the time, virtual machine, storage or network port failures would eventually get detected using software polling or timeouts, but while these long durations were acceptable in enterprise use cases, they were unacceptable in NFV. The Doctor project started to address this by working with numerous upstream OpenStack projects such as Nova, Neutron, Congress and Aodh.

The goal of OPNFV Doctor was straightforward: once an infrastructure failure is detected, map it to the affected virtual resources, and take failover or any other action within 1 second. The challenges were numerous:

- Defining what infrastructure failure is
- Mapping it to affected virtual resources
- Marking the status of affected resources appropriately
- Generating alarms to a higher-level fault management system

OPNFV Doctor may also be used with the OpenStack Vitrage root-cause-analysis project. The combined OPNFV Doctor and OpenStack Vitrage solution involves the following steps:

Fault Management Event Flow (Doctor + Vitrage)

	❶ Monitor	❷ Failure Policy	❸ RCA	❹ Update State / Notify
Physical Infra Failure				
Role	Detect and transmit raw infrastructure failure	Determine if this is a failure worth reacting to	Map failure to virtual resources; RCA visualization	Update virtual resource state; create alerts (raw & virtual)
Upstream Projects	Collectd (see OPNFV Barometer) Monasca Zabbix Nagios	OpenStack Congress	OpenStack Vitrage	OpenStack Nova OpenStack Neutron OpenStack Cinder OpenStack Aodh

Monitor

The first step, of course, is to monitor a failure in the infrastructure. Doctor allows for many different options here, including `collectd` (see OPNFV Barometer). The raw failure information is transmitted to the failure policy engine.

Failure Policy Engine

A raw failure does not imply that the failure has to be dealt with by the MANO layer. For example, if there are two redundant NICs and one fails, an operator may choose to not consider it a failure. Of course, the NIC has to be replaced, but that's a different topic. Or take an example where memory utilization reaches 80%. Some operators may consider that a failure that needs to be dealt with. Others may not. For this reason, a policy engine is needed to determine if the raw failure is indeed a failure from the MANO layer's point of view or not. OpenStack Congress, a policy-as-a-service project accomplishes this by enabling the user to create policies. The Doctor project contributed PushDriver and `DoctorServiceDriver` datasource drivers to Congress to enable the policy engine to evaluate events coming from the above-mentioned Monitor.

Root Cause Analysis (RCA) – Update State – Notify

Once the policy engine establishes that the failure is worth acting upon, multiple things needs to happen. First, the affected list of virtual machines, storage and ports has to be determined. Next, all affected physical and virtual resources have to be marked appropriately (e.g. Error, Suboptimal, Down state, and so on). Thereafter, alarm(s) have to be sent to the notification engine, and the data has to be made available to the user in an easy-to-understand visual form. The OpenStack Vitrage project does all of the above. Vitrage creates a topology graph, pulled from a variety of data sources, that builds the connections from physical to virtual resources to applications. Next the user needs to provide templates that describe how to react to events.

A sample Vitrage template is shown below (source OpenStack Barcelona Presentation):

```
- scenario:                                                              Raise alarm on
    condition: high_cpu_load_on_host and host_contains_instance          VM
    actions:
      - action:
          action_type: raise_alarm
          action_target:
            target: instance
          properties:
            alarm_name: CPU performance degradation
            severity: warning
      - action:
          action_type: set_state
          action_target:
            target: instance
          properties:
            state: SUBOPTIMAL
```

```
- scenario:
    condition: high_cpu_load_on_host and host_contains_instances
and alarm_on_instance
    actions:                                                             Add causal
      - action:                                                          relationship
          action_type: add_causal_relationship
          action_target:
            source: zabbix_alarm
            target: instance_alarm
```

```
- scenario:                                                              Set host state
    condition: high_cpu_load_on_host
```

```
actions:
  - action:
    action_type: set_state
    action_target:
      target: host
    properties:
      state: SUBOPTIMAL
```

Now Vitrage is armed to act. Once it receives an event from Congress, it will evaluate what part of the topology graph is affected and what templates need to be executed. This step "deduces" what alarms need to be sent up to the management layer. Vitrage can map the failure to virtual resources and also to affected applications. In parallel, Vitrage also marks virtual and physical resources appropriately as required by the template. Finally, Vitrage makes a variety of views available to the user for visualization:

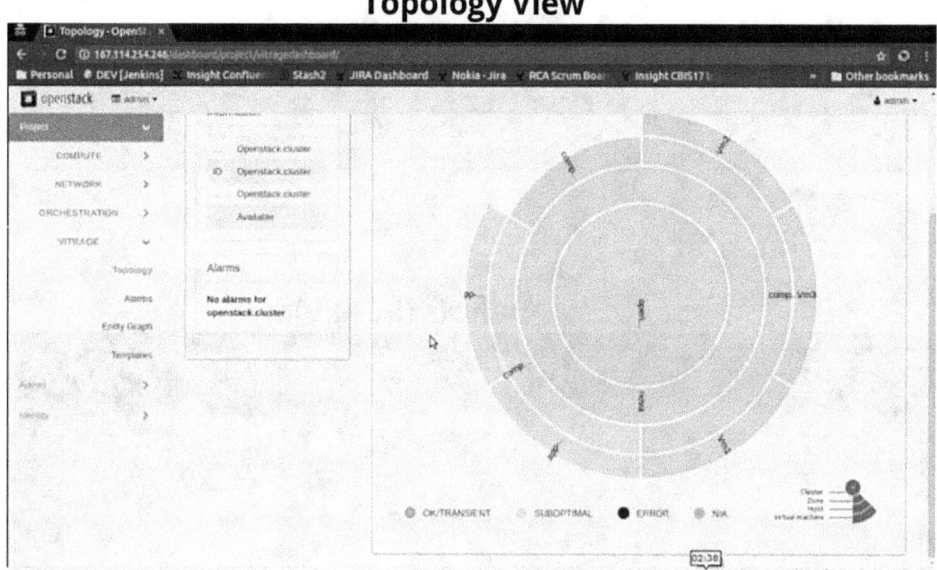

Topology Graph

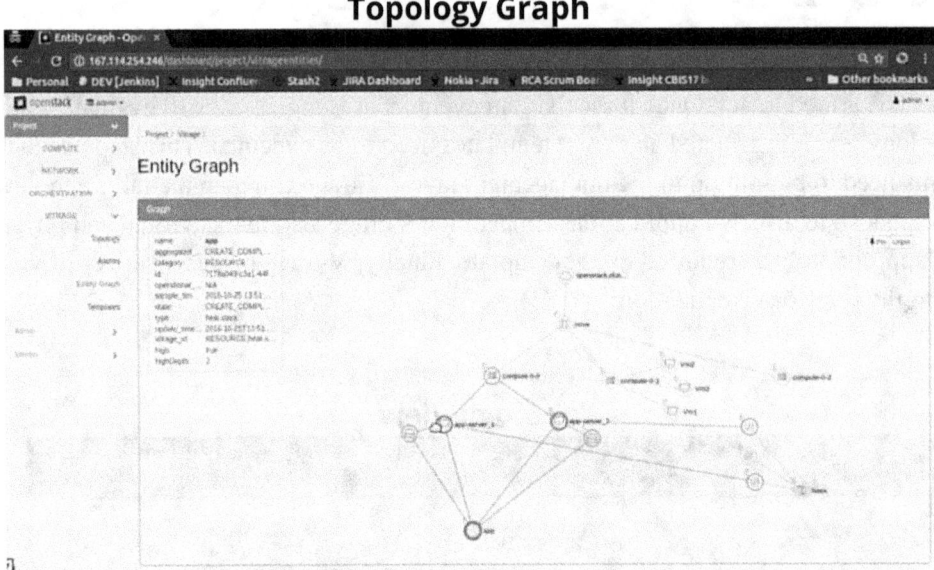

Root Cause Analysis (RCA) View

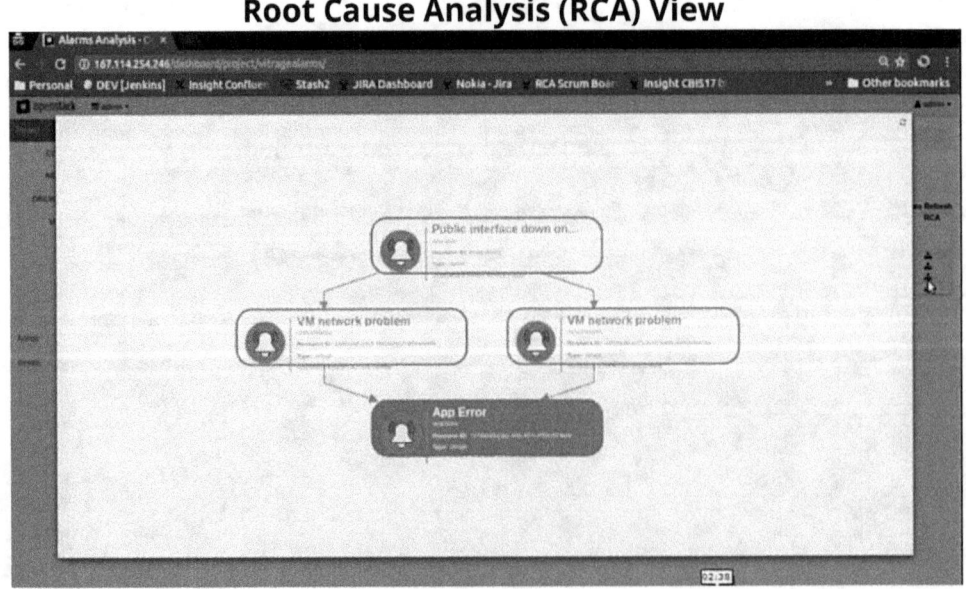

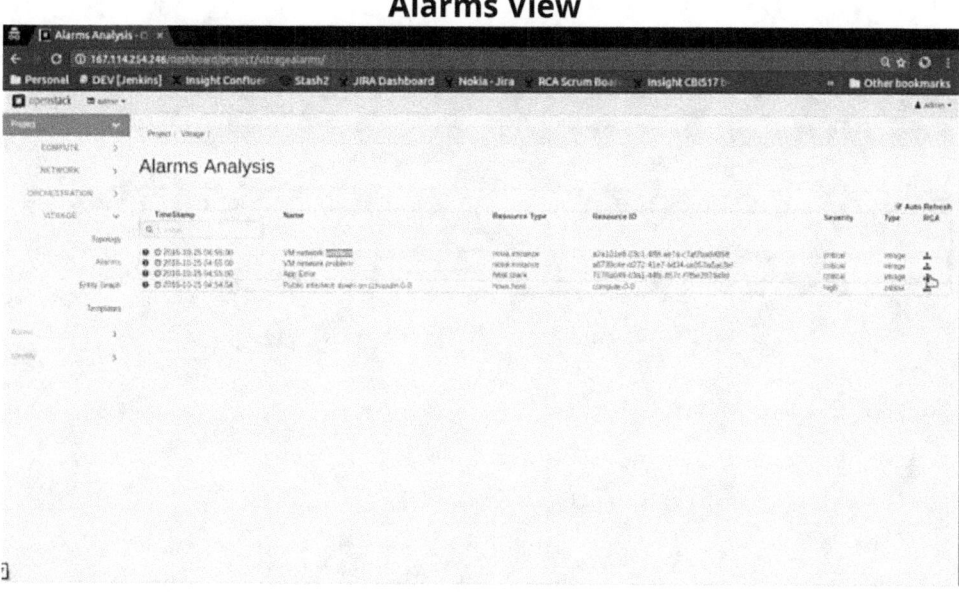

OpenStack Nova, Neutron or Cinder projects send failure notification event(s) to a software component such as OpenStack Aodh (which provides alarms and notifications) which then communicates to the MANO software. The MANO layer can then take an appropriate action, such as triggering a failover. During the OpenStack Barcelona summit, a vEPC demonstration showed that a failover in response to network failure was completed in less than 500ms, without interrupting an ongoing cell phone call. See the short video at youtu.be/Dvh8q5m9Ahk.

In summary, OPNFV upstream contribution projects are staffed by experts that are well versed in the best practices in carrier grade deployments, and they use this knowledge to recreate those same service assurance, QoS, and availability guarantees in the NFV world. In the next chapter, we will look at the next pillar of OPNFV: Continuous Integration.

6

OPNFV CONTINUOUS INTEGRATION

As we saw in Chapter 4, OPNFV integrates a number of upstream projects. To achieve this integration in an automated manner, OPNFV has created a Continuous Integration (CI) pipeline supported by three infrastructure (infra) projects[3]: RelEng, Pharos and Octopus. These three projects are critical for the success of the OPNFV project. We will review these projects and then see how they are used for integration and release processes. Chapters 7 and 8 will discuss how the CI pipeline is used for deployment and testing.

OPNFV RelEng: Release Engineering

The RelEng project defines and supports the software infrastructure required to make OPNFV a success. It collects requirements from various OPNFV projects, sets up all the tools, software automation jobs, scripts, and so on — everything needed to automate integration, deployment and testing. The project also provides guidance and support to other projects on the best practices around using the software infrastructure. The main tools provided by the RelEng along with Linux Foundation infrastructure team are as follows:

- **Collaboration:** JIRA/Confluence
- **Source code management and code review:** Git, Gerrit and Github
- **CI/software automation:** Jenkins
- **Artifact repository:** Google cloud storage and Docker hub

---

[3] These projects are used by deployment and testing projects as well

Let's discuss each tool in further detail.

Collaboration – JIRA/Confluence

OPNFV uses Confluence Wiki for community collaboration, and JIRA for bug and issue (epics and user stories) tracking and to assign issues to contributors. The combination of a Wiki page for each project, weekly meetings (open to public), JIRA stories and some additional tools described below, provides for an open and transparent way for any individual to contribute.

OPNFV JIRA

Source Code Management and Code Review

OPNFV uses three tools for source code management and code reviews: Git, Gerrit and Github.

Source Code Management – Git

Code, scripts, template files and source for documentation that is automatically generated for all OPNFV projects are kept in Git repositories. Git is a distributed open source revision control system originally developed by Linus Torvalds to maintain the golden repository of a project (trunk) and manage all the branches (forks). Developers can create a branch from trunk, work on their patch, and then submit their changes to be committed to trunk.

An example OPNFV Git repository

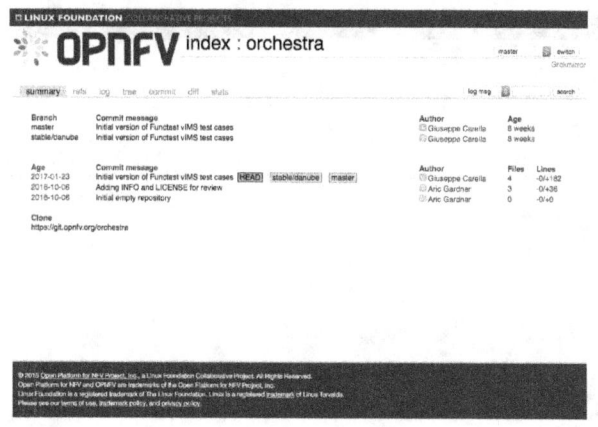

Code Reviews – Gerrit

Committing to master requires an approval process, and this process is managed through a tool called Gerrit. Gerrit is an open source web-based code review tool developed by Google. All changes pushed by contributors using a `git push` or `git review` command are reviewed in Gerrit by a set of reviewers, who view and inspect the patch. Reviewers also get to see the results of a continuous integration (CI) build and automated `verify` test run. Reviewers provide scores of +2, +1, -1 or -2. A +2 is a definite accept, while a -2 is a definite reject. A +1 or -1 may result in the change being accepted, rejected or sent back for changes.

OPNFV Gerrit

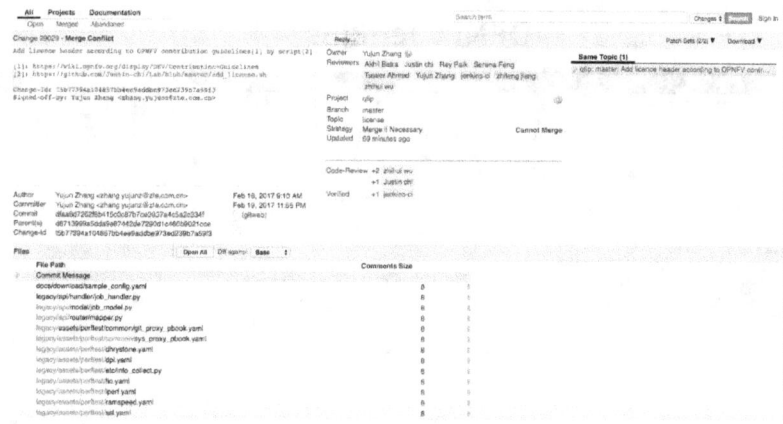

Source Control Management Mirror – Github

OPNFV uses Github for mirroring its Git repositories.

An example OPNFV Github mirror repository

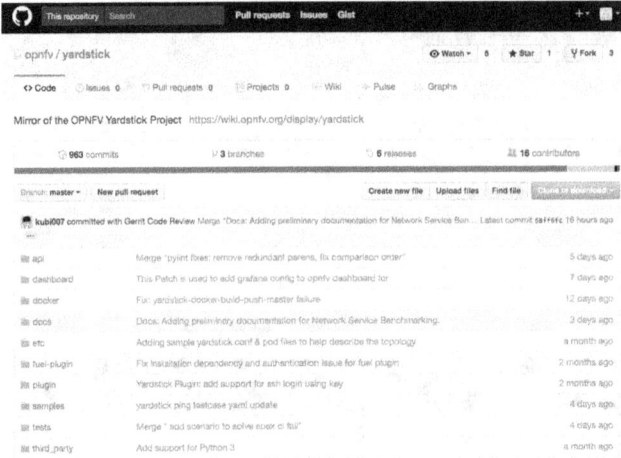

Software Automation/CI – Jenkins

Jenkins is an open source software automation tool that is commonly used to automate integration builds and testing. Think of Jenkins as an amalgamation of powerful scripting, a `make` utility on steroids, and the ability to trigger activities based on events such as Git repository changes. OPNFV uses Jenkins to run automated continuous integration, deployment and testing. The tests may be simple `verify` jobs or one of the more complex test suites (see Chapter 8). OPNFV treats Jenkins jobs like source code; these files are developed, reviewed, tested and submitted. Jenkins Job Builder (JJB) enables this process by automating the creation, testing, configuration, and maintenance of the Jenkins jobs used by OPNFV CI.

OPNFV Jenkins

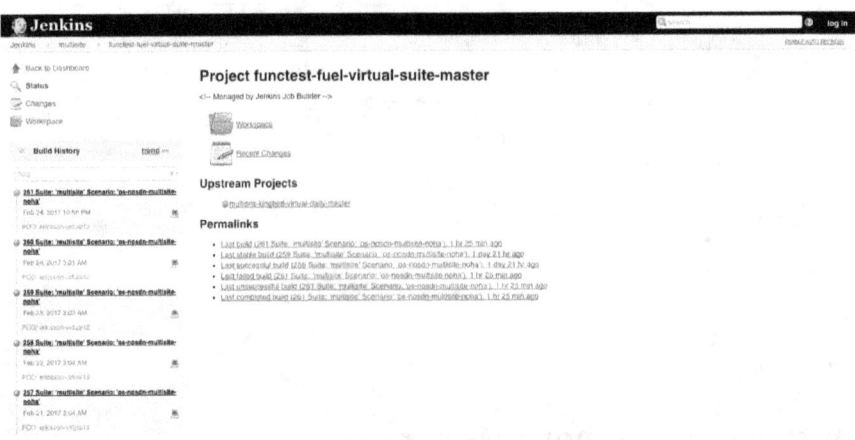

RelEng also provides additional tools for test result data collection, analytics, search and visualization, such as Elasticsearch, Logstash, Kibana and Grafana.

Artifact Repository – Google Cloud and Docker Hub

Git is great for source files, but it is not a good place to store large artifacts such as detailed documentation, Docker images, packages, ISO files, and so on. These are stored in an artifact repository currently hosted on Google Cloud Storage, or in the case of Docker images, on Docker hub.

OPNFV Pharos: Community Lab Infrastructure

In addition to software infrastructure, OPNFV projects need a corresponding hardware infrastructure project for development, integration, deployment and testing. That's the goal of the Pharos project. Pharos develops a geographically distributed, heterogeneous, community lab infrastructure using bare metal servers. Currently, there are 16 labs hosted by the Linux Foundation and individual companies across North America, Europe and Asia. The diversity of labs in terms of vendors, geography, and CPU architectures is a major asset towards the hardening of OPNFV software.

The rigor of testing on Pharos is impressive; for any particular release, Pharos supports the creation and teardown of thousands of OpenStack + SDN controller clusters with multiple tests suites run against them. This is actually superior to the testing performed by corresponding upstream communities. It's no wonder that upstream communities get tremendous value out of OPNFV testing.

CENGN Pharos Lab[4]

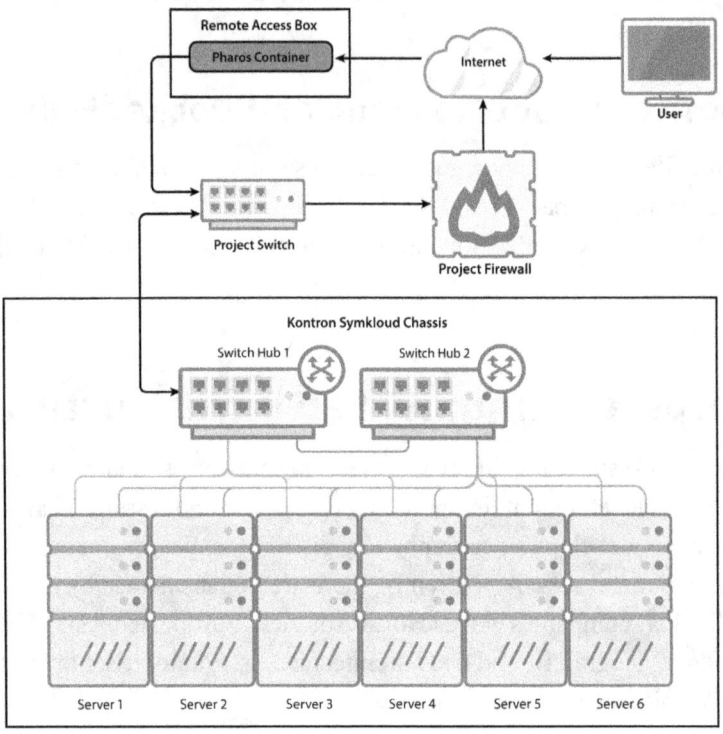

The Pharos project has multiple activities:

- **Specification:** The specification activity lays out minimum requirements for each individual environment or POD. A POD is a jump server with a set of five servers used as controllers, compute and storage nodes. A POD also needs a number of networks and lays out storage and switching requirements. One or more PODs make up a lab. The lab provider needs to publish specification files describing the lab and each POD. The *specification* work stream continues to build additional specifications for security, stability and lifecycle management.
- **Dashboard:** The Pharos dashboard provides a holistic view of all labs, resources and utilization.

---

[4] Shown with permission

OPNFV Pharos Dashboard

- **Lab-as-a-service (LaaS):** While the key goal of Pharos is testing, some projects need access to bare metal resources for development purposes. LaaS provides lab resources to projects in the form of a service.
- **Validator:** These are a set of tools to validate that a lab is compliant with the Pharos specification.
- **Virtual labs:** Most of the focus is on bare metal resources. But for development, virtual labs will also be made available.

OPNFV Octopus: Continuous Integration Project

The Octopus and RelEng projects create the CI pipeline described below, which deploys and tests OPNFV software on PODs in various Pharos labs. The pipeline is rich in functionality, with jobs such as build, verify, merge, test, and so on.

CI Pipeline

The pipeline for build, test and release looks like this:

Simplified CI Pipeline Diagram

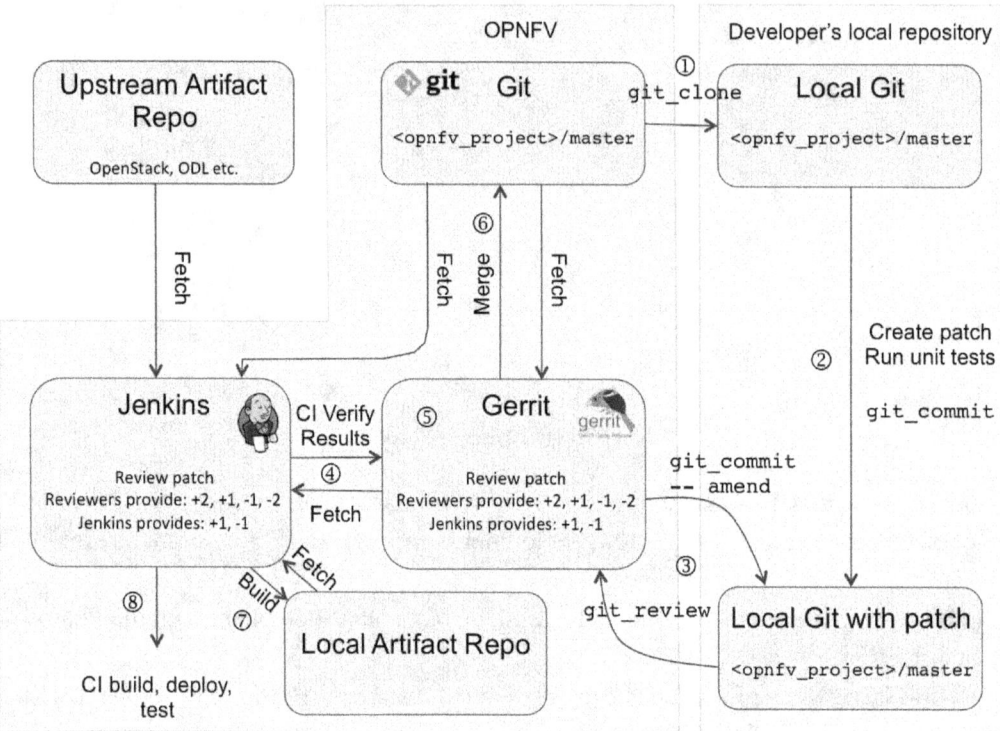

1. A contributor clones git master to a local repo.
2. The contributor makes changes, adds code or fixes bugs and performs local unit testing.
3. The contributor submits the patch for review.
4. The patch is verified using a Jenkins CI `verify` job.
5. Reviewers can inspect the patch and the CI result; the patch may be accepted, rejected or sent back for revisions.
6. If the patch is accepted, one of the committers submits the patch to Git trunk, which triggers a Jenkins `post-merge` job.
7. Jenkins jobs also create local artifacts, such as Docker containers for test tooling.
8. Finally, Jenkins jobs run a variety of tests on a daily, weekly, or non-recurring basis. These tests run on OPNFV software deployed using code built from the Git trunk, upstream artifact repositories (such as OpenStack, ODL etc.) and local artifact repositories (such as Docker containers for test tooling).

What About Releases?

At this time, OPNFV major releases are on a regular six-month cadence. Creating releases on a regular rhythm is a common practice amongst open source projects. Close to the release date, the focus shifts from adding new functionality to bug fixes. The release team also makes decisions on whether to include or exclude specific projects based on maturity and readiness.

As you might expect, the ultimate goal is to evolve to a continuous delivery approach where releases are available at a higher frequency and ultimately, the trunk is stable at any given point in time.

Making CI/CD More Continuous

The use of DevOps and CI is quite advanced in OPNFV; however, there's always more to do.

One important aspect is security checking, where every commit will get checked for security violations using automated tooling. Under consideration are checks such as Lint and vulnerability scanner for source code files, the inclusion of binary objects, use of strings that might contain passwords, keys or hashes and missing licenses.

When it comes to consuming upstream projects, the default practice is to use regular stable releases. The next step, already in place with certain projects, is to consume upstream projects on a more regular basis, even daily. Using this process, OPNFV can do two things i) provide feedback for each commit using health checks and ii) run daily and weekly tests on the latest code base. The OPNFV Cross Community CI initiative, or XCI, covers these newer processes initially for OpenStack, ODL, FD.io and ONAP. The benefits of XCI are faster access to the latest innovation and rapid feedback to the upstream community.

CI/CD Today & in The Future

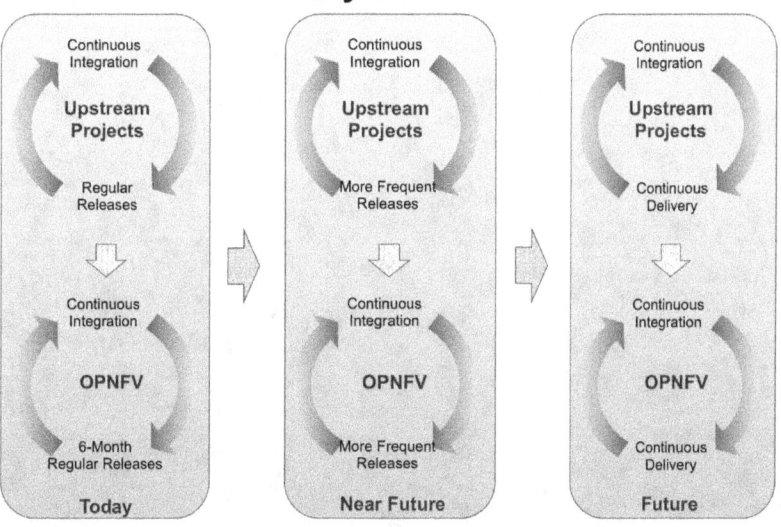

These improvements are some of the more visible, but constant evolution is going on behind the scenes as well, in an attempt to be cleverer in the execution of tests, to improve the utilization of resources, and to help developers gain more insight and feedback.

In summary, OPNFV integrates upstream projects for the purpose of running different tests. To achieve this, OPNFV uses sophisticated continuous integration to ensure access to the latest innovation. So far, we have not talked about the exact mechanism of how testing is actually performed. In the next two chapters, we will see how the upstream projects are deployed, and how tests are run against those deployments.

7

OPNFV AUTOMATED SOFTWARE DEPLOYMENT

One of the main goals of OPNFV CI, as we saw in the previous chapter, is the continuous and automated testing. The testing process occurs in two steps:

1. The CI process deploys a set of upstream projects with specific configurations
2. The CI process runs automated test(s) against the deployed software

In this chapter, we will cover the first step; we'll look at the second step in Chapter 8.

What Are OPNFV Scenarios?

An obvious question emerges: if the OPNFV project integrates multiple software projects for several layers, then what exactly is being tested? For example, you can't possibly test ODL, ONOS and OpenContrail all at once. Nor can you test OVS and FD.io at once. Also, there are numerous variations in configurations, such as high availability, DPDK, service function chaining (SFC), BGPVPN and so on, where choices have to be made. To solve this conundrum, OPNFV has come up with the concept of *scenarios*.

> Scenarios are a deployment of a set of components and their configuration

Scenarios are a particular combination of components, along with their configuration. It is impractical to try out every possible combination, so the OPNFV community deploys and tests the highest priority scenarios, based on community interest and end user feedback (e.g. from the End User Advisory Group – EUAG). For the Danube release, there are 55 scenarios. This versatility of OPNFV provides users with the confidence of choosing from a number of pre-tested combinations of upstream projects that suit their requirements. The number of scenarios has been climbing over time, as shown below.

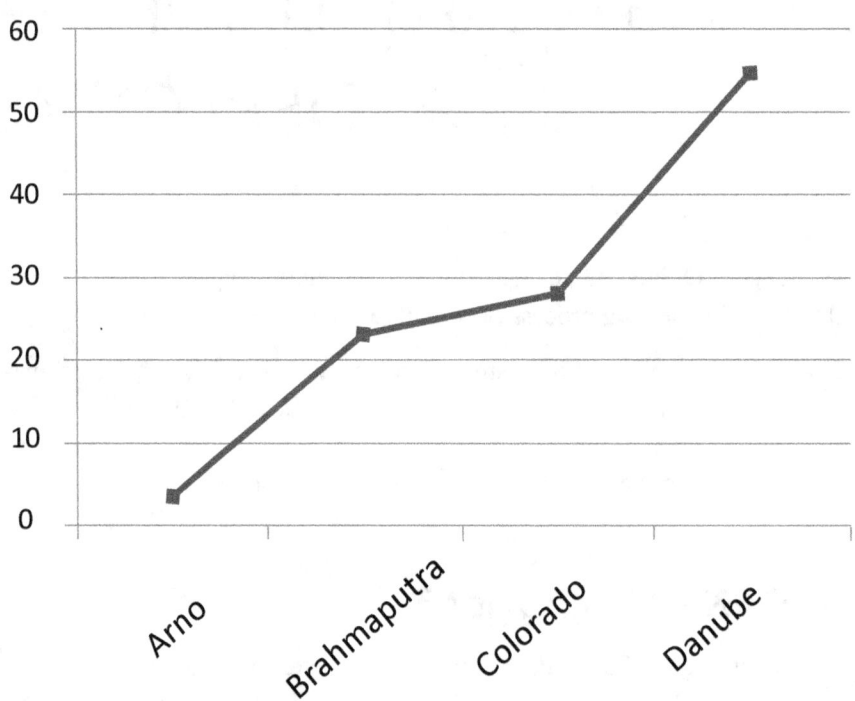

In addition to providing choices for different layers of the NFV stack, OPNFV also provides choice in the installer that deploys and configures the software. Diving a bit deeper, scenarios are labelled using the below notation. So, using the below key, `os-odl_l2-fdio-ha,Fuel` means: OpenStack with ODL Layer2 SDN Controller and FD.io in High Availability mode installed by Fuel.

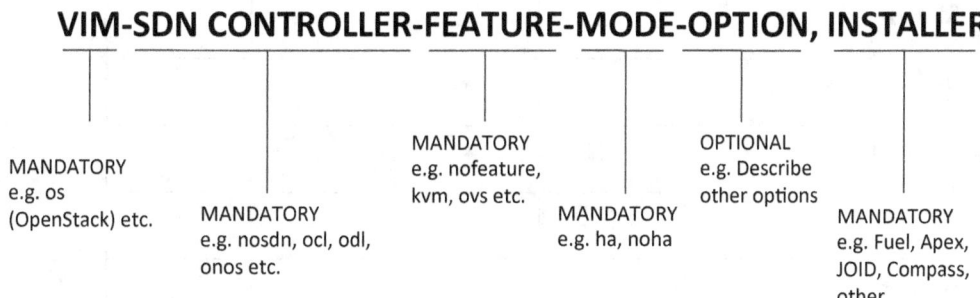

However, not all 55 Danube scenarios qualify as reference architectures that could be deployed to a production environment. Twenty of them do not use high availability (i.e. noha), and 3 more are experimental because they use Kubernetes as the VIM or Daisy as an installer. So now you are down to 32, as shown below.

	Fuel	Apex	JOID	Compass
os-nosdn-fdio-ha		x		
os-nosdn-kvm_ovs_dpdk_bar-ha	x			
os-nosdn-kvm_ovs_dpdk-ha	x			
os-nosdn-kvm-ha	x	x		
os-nosdn-lxd-ha			x	
os-nosdn-nofeature-ha	x	x	x	x
os-nosdn-openo-ha				x
os-nosdn-ovs-ha		x		
os-ocl-nofeature-ha			x	
os-odl_l2-bgpvpn-ha	x			
os-odl_l2-fdio-ha		x		
os-odl_l2-nofeature-ha	x		x	x
os-odl_l2-sfc-ha	x	x		

os-odl_l3-fdio_dvr-ha		x		
os-odl_l3-fdio-ha		x		
os-odl_l3-nofeature-ha	x	x		x
os-odl_l3-ovs-ha		x		
os-odl-bgpvpn-ha		x		
os-onos-nofeature-ha		x	x	x
os-onos-sfc-ha			x	
os-ovn-nofeature-ha		x		

As you can see, if we ignore the association of scenarios to a specific installer, we are down to 21 scenarios that could be used as candidates for creating production environments.

Installers

In the spirit of providing choices, OPNFV uses four installers. These installers deploy all the software required for a scenario, and then also configure the software. Even though there are many differences, the fundamental mechanism of every installer is similar: the installer first installs an application on a specified "jumphost", which in turn runs and downloads, installs and configures all of the necessary software on designated compute and controller nodes resulting in a fully deployed scenario. This sets the stage for tests to be run against the scenario. The use of installers is fully automated through the CI pipeline.

Frankly, if the only job of the installer was to set up these scenarios, there is no need for using more than one installer. However, these installers also have a life outside of CI testing. Beyond installation, they can also be used for day-2 lifecycle management tasks such as post-deployment configuration changes, capacity expansion, functionality enhancements, updates and upgrades. Some installers also have integration with monitoring software. For this reason, OPNFV provides installer choices. At this time, OPNFV does not test any day-2 functionality, so while interesting, a discussion about those capabilities is outside the scope of this book. Let's take a deeper look at the 4 installers from an initial deployment (day-1) point of view.

Fuel

Fuel is a purpose-built OpenStack installer community project that came out of Mirantis.

Sample Fuel Screenshot

Fuel supports APIs, a CLI, or a GUI. Once the Fuel master node starts up, all compute, controller and storage nodes are discovered using PXE. Configuring the cloud is a simple matter of point-and-click by going through different tabs organized by functionality. There is a network check to ensure that the environment is correctly configured before the software is deployed. After the deployment is completed, there is a post-deployment health check as well. Fuel has an extensible framework via plugins that allow for new functionality to be integrated into Fuel (for example, plugins for an SDN controller, MANO, and so on).

Apex

Apex is RDO's OpenStack installation tool, based on the OpenStack Triple-O project.

(RDO is a project that takes OpenStack packages originally developed for Ubuntu and ports them to CentOS, Fedora and Red Hat Enterprise Linux.) Triple-O uses an interesting philosophy of OpenStack on OpenStack. It first installs a light-weight OpenStack "undercloud". An OpenStack "overcloud", the one that will actually be used by workloads, is deployed by the undercloud using Heat templates. Apex provides flexibility, but at the cost of numerous manual steps.

JOID

JOID stands for Juju OPNFV Infrastructure Deployer. JOID uses two open source projects that came out of Canonical: MAAS and Juju. Both are general purpose technologies, as opposed to purpose-built.

Sample JOID Juju Screenshot

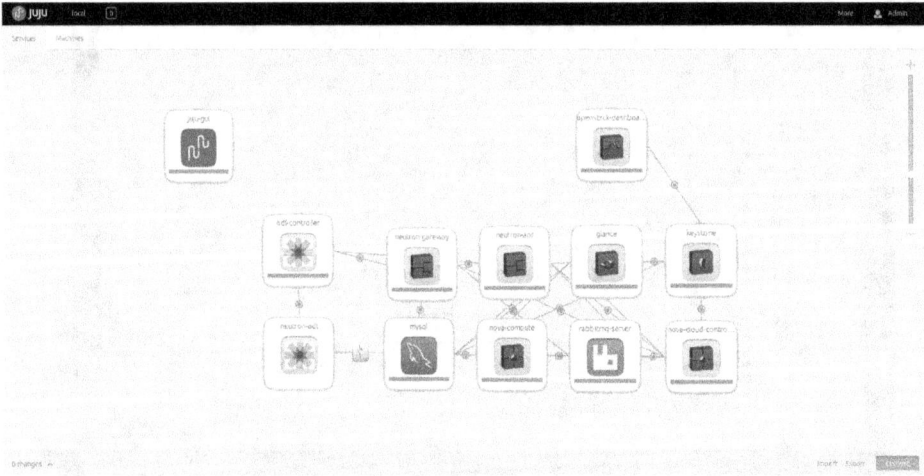

MAAS (metal as a service) is a tool for provisioning bare metal servers. MAAS has functionality such as OS provisioning, inventory management, IPAM (IP address manager) and PXE/IPMI integration with a convenient GUI, CLI, or API access.

Juju is a service orchestration tool that is used for deploying OpenStack on a MAAS environment. If Juju sounds familiar, it's because we discussed it in Chapter 4 in the context of being a VNFM for multiple MANO projects. Juju uses descriptors called Charms that are a template file to define a service. Charms support languages such as Chef, Puppet, Ansible and so on, and allow for the easy addition of new functionality such as SDN controllers or MANO.

Compass

Compass is another OpenStack community purpose built installer project, from Huawei.

Sample Compass Screenshot

The tool can be accessed via a GUI, API or CLI. It has a wizard-like feel where Compass systematically moves the user across various stages of deployment. It starts with discovery, which can use a variety of mechanisms such as IPMI, SNMP and Intel® RSA. The next stage is provisioning of the OS – Centos and Ubuntu are both supported. After a number of configuration steps (network, storage partitioning, security, role assignments for various roles, and network mapping), various OpenStack software packages are deployed using Chef or Ansible. Compass also integrates infrastructure monitoring dashboards for logs and metrics.

Additional Software Deployment Projects

In addition to the four installers, OPNFV includes additional software deployment projects:

Project	Description
Daisy (Installer using	The future of OpenStack deployment seems to be to

containerized control plane)	containerize OpenStack services and orchestrate those containers. Doing this dramatically improves the day-2 experience. ZTE's open source Daisy project uses OpenStack Kolla, a project that provides production-ready containers for OpenStack services, and then builds on it with tooling for automated deployment, configuration templates etc. Daisy4NFV is still in incubation, and promises to ease deployments for large scale clouds.
IPv6-enabled OPNFV	The project, as the name suggests, identifies gaps in upstream projects to implement a full IPv6 NFV deployment and also creates IPv6 scenarios. For the Danube release, the project can use Fuel, JOID and Compass.
Deployment Template Translation (Parser)	The VNFM layer and VNF descriptors provide NFVI requirements (such as vCPU, memory, storage, dataplane acceleration, and so on) and specify post-deployment records (such as utilization, performance report, and so on) in formats such as TOSCA. But these templates often need to be translated to another format such as a Heat template for the VIM layer to act upon them. This project solves the translation problem through four translator projects: `tosca2heat`, `yang2tosca`, `policy2tosca`, `tosca2kube`.
ARMBand	The goal of ARMBand is to make sure OPNFV software runs on ARM. It replicates all aspects of OPNFV, such as software build, CI, lab provisioning, testing processes for the ARM architecture. At this time, a subset of scenarios and test projects are supported via ARMBand.
FastDataStacks (FDS)	The FDS project uses APEX to deploy FD.io scenarios.

Currently installers are tied to particular PODs. To improve POD utilization and for projects such as Multisite (see Chapter 5) it becomes important to decouple installers from PODs, and the future direction of OPNFV is to make the mapping of scenarios to POD(s) a lot more flexible and dynamic.

In the next chapter, we will look at how tests are run against each scenario in an automated and continuous manner.

Understanding OPNFV

8

OPNFV CONTINUOUS TESTING

All OPNFV activities discussed so far build up to the finale – continuous testing. It is important to reiterate how critical end-to-end testing of the full stack is, and that OPNFV is the only community currently driving this. Through this testing, the OPNFV community is able to find bugs and security issues that cannot be discovered by testing each component individually.

The test projects in OPNFV are meant to validate the numerous scenarios using multiple test cases. The following figure shows a high-level view of the OPNFV testing framework. The rest of the chapter will analyze this figure in more detail.

OPNFV Testing Community

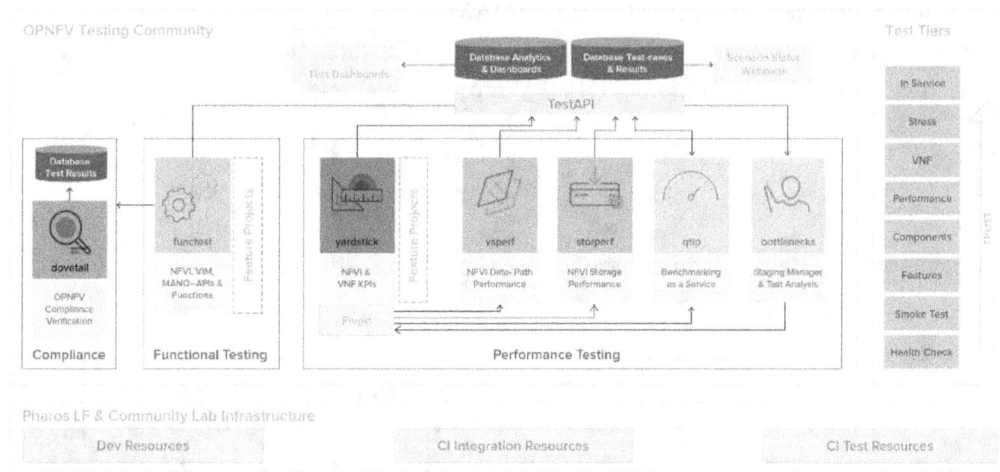

Once a scenario is deployed by the installer, a variety of test cases are run against that scenario. These tests vary in terms of coverage, scope, tiers, purpose and frequency of testing. Both the test cases and the results are stored in a database. The database is connected to a number of visualization tools to view the results. As we have talked extensively, all these tests are part of the CI pipeline, and are run in a completely automated manner.

Classification of Test Cases

OPNFV test cases may be classified in a number of different ways: by coverage, scope, tier and purpose:

Coverage

The coverage of a test case can be end-to-end, be restricted to a single component or to a subsystem. Components tests cover individual items of the NFV stack such as OpenStack, NICs, ODL, VNFs, and so on. Subsystem (also called ETSI domains) tests can be the entire VIM, the storage aspect of the NFVI, a complete service consisting of VNFs (such as vIMS), and so on. End-to-end test cases cover all components of an entire scenario as a cohesive unit.

Scope

The scope of each test case is split into three broad categories: functional, performance and compliance. For now, there aren't enough stability or security tests in OPNFV, so these are bucketed with functional testing. As the emphasis on these areas increases, they could be broken out into their own categories. Test cases that cover OAM (operations, administration, management – including lifecycle management), longevity (how long an environment is stable), scale, and destructive testing are not in the OPNFV scope at this time.

Tiers

OPNFV has numerous test tiers that increase in complexity and reduce in frequency.

Tier	Example	Frequency	Trust
Other	Bottleneck & Qtip projects (see below)	Non-recurring	High

VNF	vIMS (Clearwater) test	Weekly	↑
Performance	Full performance testing	Weekly	
Components	Full OpenStack testing, specific performance tests	Daily	
Features	Specific project tests, such as Doctor, Promise, security scan, HA, IPv6, SFC, and so on	Daily	
Smoke	Basic smoke, sanity, vPing tests	Daily	
Verify (Health Check)	Gating test	Commit test	Low

A trust indicator has been defined for test cases and scenarios. Trust simply means that with each subsequent tier, the confidence in the scenario being tested increases. Test cases with higher levels of trust are also run at a lower frequency, because they are often long running tests, and hardware resources are always limited. In this manner, the trust indicator helps both users in evaluating scenarios and the infrastructure team in optimizing CI resources.

Purpose

A test case may have one of many purposes:

- **To gate commits:** The verify or health-check test is run in response to a commit to provide a +1 or -1 to gerrit.
- **To gate a release:** These are tests that *must* pass for a release to occur. As mentioned earlier, major OPNFV releases are on a 6-month cadence. Each major release also has minor releases, such as Danube 1.0, 2.0 and 3.0, each of which share the same stable branch and vary only in terms of bug fixes.
- **Informational only:** A number of performance tests do not gate a release. They are simply run to publish informational data.
- **Benchmark:** These metrics synthesize a benchmark based on performance metrics.
- **OPNFV compliance:** These tests test against a set of validation criteria to be able to use OPNFV trademarks. They are not to be confused with regulatory compliance testing.

One additional point: each individual upstream contribution project is responsible for developing its own tests, and global tests that are not under the purview of a given project are covered by the activities of test projects. Let's take a deeper look into those test projects.

Test projects

OPNFV test projects are organized by scope. At the top level, there are functional tests (Functest), followed by performance tests (Yardstick along with a number of sub-projects) and OPNFV compliance tests (Dovetail).

Functest

The Functest project deals with the tooling and actual test cases around validation and functional testing of various OPNFV scenarios. The current focus is around VIM + NFVI functionality. The baseline is to use test cases from various individual upstream projects, such as Tempest (functional) and Rally (scale) for OpenStack, the Robot framework for ODL, and the Teston framework for ONOS. There are four reasons to repeat these upstream tests in Functest: 1) it tests projects together to ensure end-to-end interoperability, 2) it adds functional tests for the numerous OPNFV upstream contribution projects (the tests are also contributed back to the upstream projects), 3) it adds a number of end-to-end test cases, and probably the most important, 4) it incorporates open source VNFs, such as Clearwater vIMS, along with MANO projects to generate real-life test conditions on the underlying VIM + NFVI layers.

Yardstick

Yardstick is all about performance testing and is based on ETSI reference test suites. After surveying a large number of NFV workloads, the Yardstick project has broken down overall requirements into a set of performance vectors to quantify compute, network and storage aspects of NFVI. The project develops the test framework and test cases for each performance vector, and some test cases can get quite complex where Yardstick needs to run tests in parallel, inject faults, and test multiple topologies.

Yardstick is extensible via a plugin architecture. It exposes and consumes APIs from specific performance testing sub-projects such as VSPERF (vSwitch), Cperf (SDN controller), Storperf (storage), Qtip (Benchmarking as a service) and Bottlenecks (bottleneck detection).

Yardstick also includes traffic generation tools, as well as a Network Service benchmarking module in order to evaluate performance for onboarded VNFs.

VSPERF

The virtual switch (such as OVS) is a major aspect of a VNF's performance, as measured by throughput, jitter, packets per second, and processing latency. For this reason, the VSPERF project measures the performance of the vSwitch and associated virtual and physical network ports. The project currently evaluates the performance of OVS, with and without DPDK, though the test project is agnostic of both the vSwitch implementation and the traffic generator. Performance is measured for paths such as:

- Port → vSwitch → Port
- Port → vSwitch → VNF → vSwitch → Port
- Port → vSwitch → VNF → vSwitch → VNF → vSwitch → Port

Different test cases, based on a number of industry specifications and use cases, measure a variety of aspects such as forwarding rates, the impact of a noisy neighbor, datapath and control path coupling, CPU and memory utilization, and so on. VSPERF tests a vSwitch as if it were a physical switch and uses an external traffic generator, and is careful to ensure accuracy, consistency, stability and repeatability of tests across runs. VSPERF can be launched in standalone mode or through Yardstick.

Cperf

Cperf measures the performance of the SDN controller. It taps into the expertise of the upstream project's performance teams, such as the OpenDaylight Performance Group. The project runs a variety of performance tests, such as:

- Network scalability (for example, max switches, ports, links, and so on) with one controller node
- Cluster scalability (for example, max controllers)
- Network scalability (for example, max switches, ports, links, and so on) with a cluster of controller nodes
- Flow performance (for example, max flows/sec, packet latency, and so on)
- API performance (for example, Northbound, Southbound API latency, and so on)
- Datastore performance (for example, max reads/sec, writes/sec, and so on)

Storperf

Storperf measures the performance of external block storage. The goal of this project is to provide a report based on SNIA's (Storage Networking Industry Association) Performance Test Specification. The project measures latency, throughput, and IOPS (IO per second) across different block sizes and queue depths (number of outstanding I/Os in flight).

Getting down to actual steps, a Docker container with the Storperf test APIs is invoked on the jumphost. Automated tests use these APIs to spin up volumes and VMs, connect them, run a variety of storage tests, and collect the results. Storperf is suitable for both HDD and SSD storage. The methodology of using a Docker container for test tooling is common across test projects. Finally, Storperf can also be launched in standalone mode or through Yardstick.

Qtip

Remember benchmarks such as MIPS or TPC-C, which attempted to provide a measure of infrastructure performance through one *single* number? Qtip attempts to do the same for NVFI compute (storage and networking part of roadmap) performance. Qtip is a Yardstick plugin that collects metrics from a number of tests selected from five different categories: integer, floating point, memory, deep packet inspection and cipher speeds. These numbers are crunched to produce a single Qtip benchmark. The baseline is 2,500, and bigger is better! In that sense one of the goal of Qtip is to make Yardstick results very easy to consume.

Bottlenecks

Wouldn't it be great to find performance system limitations (in other words, bottlenecks) in a staging environment rather than a production environment? That's exactly what the Bottlenecks project does. Bottlenecks is integrated with Yardstick. As opposed to Qtip, where the goal is to create a new benchmark, here the goal is to use a variety of existing benchmarks and metrics to measure whether the network, storage, compute, middleware and app performance meets a user's requirements or not. The entire process is driven off "experiment configuration files" that are set up by the user. Bottlenecks drives its activity based on these files and fully automates setting up the infrastructure, creating workloads, running tests, and collecting results. The data collected from these tests tends to be quite large, so the project also invests in analytics and visualization tools. The results help identify the metric(s) that do not meet requirements, in turn enabling the user to make decisions such as hardware selection or software tuning, protocol selection, and so on, and the results assist in evaluating compliance with SLAs.

Dovetail

As we saw in Chapter 3, one of the primary benefits of using open source software is to avoid vendor lockin. This is easier said than done. How might a user verify that a vendor's distribution is consistent with the original project, and that some aspects haven't been (inadvertently) changed? In the case of OPNFV, it's even harder, because there's isn't one single distribution to deal with. Instead, there are numerous scenarios, any one of which could be considered as the basis of a distribution. Dovetail solves this problem by providing compliance tests. By running these tests, which heavily leverage Functest and Yardstick, Dovetail forms the foundation of the Compliance Verification Program (CVP) for commercial products. The goals of the CVP is to help build the market for OPNFV-based infrastructure and applications designed to run on that infrastructure, reduce adoption risks for end-users, decrease testing costs and enhance interoperability.

Dashboards

There are many dashboards to interact with and to visualize test results. We can classify them by: Scenario status reporting, complex results display (Grafana) or API interaction (Swagger). For the Danube release, they can be accessed from a central portal.

Scenario reporting results are available for Functest, Yardstick, Storperf and VSPERF. For Functest there's also reporting on the vIMS VNF testing. For example:

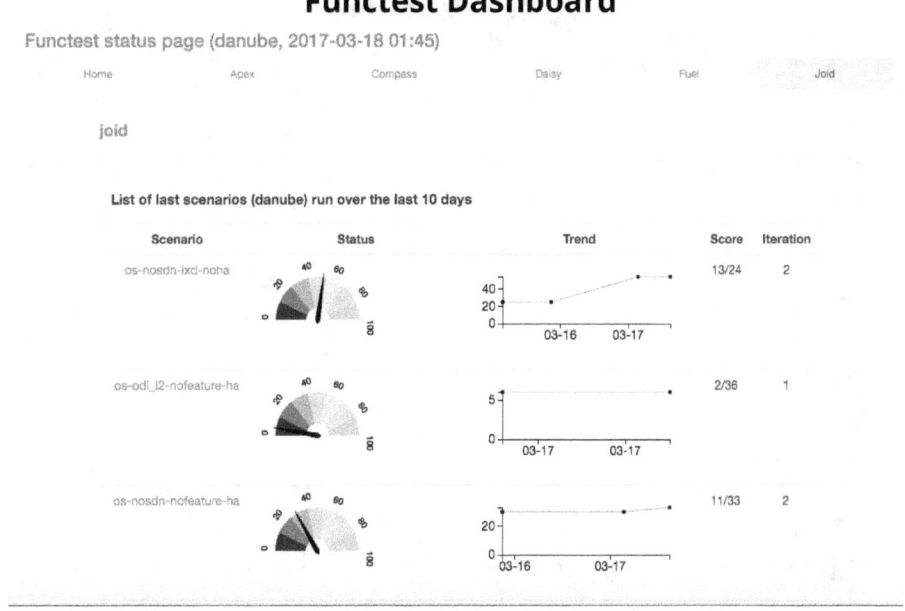

Yardstick results can be visualized through a Grafana dashboard (creds opnfv/opnfv). Any test

Understanding OPNFV

case can be viewed by scenario and POD. Since the data is coming from a time-series database InfluxDB, the user can also specify the time-period.

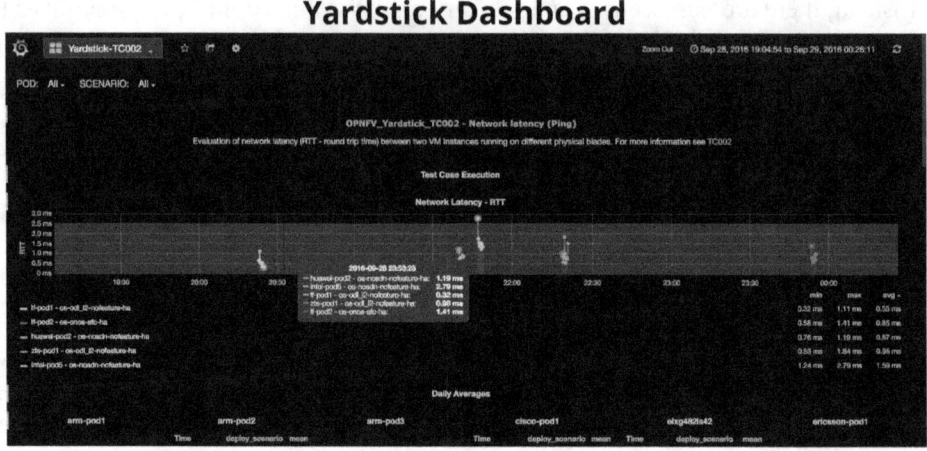

The Swagger dashboard (creds opnfv/api@opnfv) enables users to visualize and interact with the test API[5]. For example, if you try the `GET /api/v1/projects/{project_name}/cases` and type in `promise`, you will get a list of all test cases related to the project Promise.

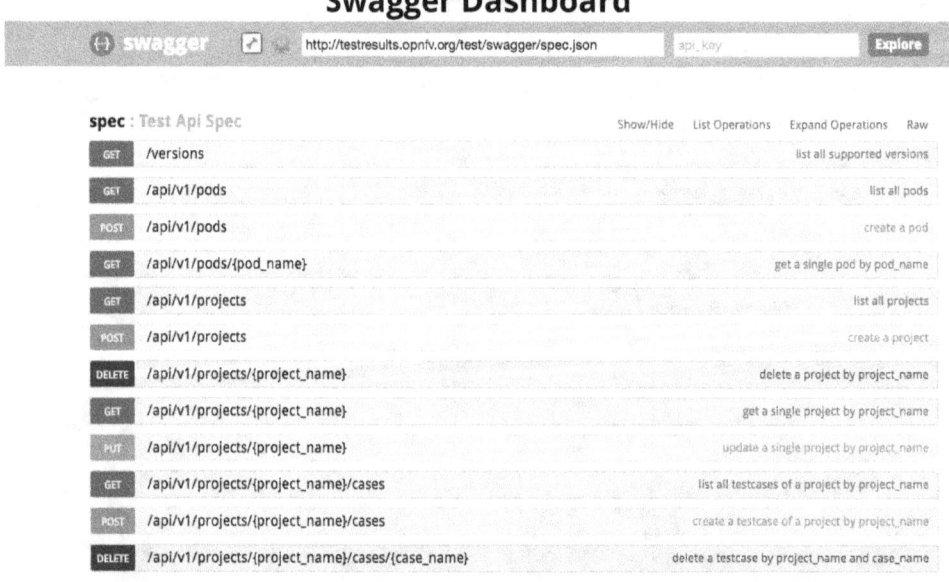

---

[5] A side note, test projects are not required to use the API framework, but most do.

Plugfests

OPNFV conducts plugfests after each release. Vendors and community members get together and collaborate on interop testing and different test projects. The plugfest serves two purposes. First, it accelerates projects significantly by having a burst of intense activity with the ability to work on issues and solve problems with other contributors face-to-face rather than remotely. Second, vendors with proprietary products that cannot be included in the OPNFV CI pipeline can test their wares across different scenarios and test cases. These plugfests are generally held concurrently with hackfests, as mentioned in Chapter 5.

In addition to the already mentioned valuable contribution of RelEng, Pharos and Octopus, there is a lot of background infrastructure work in terms of developing tests, troubleshooting failures, creating and managing databases, providing the Docker containerization framework for test project tooling, building the tooling to promote scenarios through the tiers of tests, and finally analyzing the contribution of specific scenarios and pods to overall testing results.

Finally, the OPNFV community recognizes the value of running functional and performance tests using real VNF workloads. In addition to project Clearwater mentioned above, a new project called Samplevnf has been created to do performance testing, benchmarking and characterization using five open source VNFs. These are "open source approximations" of carrier grade VNFs and not production grade; nevertheless, they are adequate for testing purposes. They are:

- CG-NAT (carrier grade network address translation) VNF
- Firewall (vFW) VNF
- Provider edge router (vPE) VNF
- Access control list (vACL) VNF
- Next Generation Infrastructure Core (NGIC) aka VEPC-SAE-GW VNF

This chapter completes the discussion of the various OPNFV projects. In the next chapter, we will discuss how VNFs can be onboarded into OPNFV.

Understanding OPNFV

9

WRITING VNFS FOR OPNFV

The entire OPNFV stack, ultimately, serves one purpose: to run virtual network functions (VNFs) that in turn constitute network services. We will look at two major considerations: how to write VNFs, and how to onboard them. We'll conclude by analyzing how a vIMS VNF, Clearwater, has been onboarded by OPNFV.

Writing VNFs

We looked at three types of VNF architectures in Chapter 2: cloud hosted, cloud optimized, and cloud native. As a VNF creator or a buyer, your first consideration is to pick the architecture.

Physical network functions that are simply converted into a VNF without any optimizations are likely to be cloud hosted. Cloud hosted applications are monolithic and generally stateful. These VNFs require a large team that may or may not be using an agile development methodology. These applications are also dependent on the underlying infrastructure to provide high availability, and typically cannot be scaled out or in. In some cases, these VNFs may also need manual configuration.

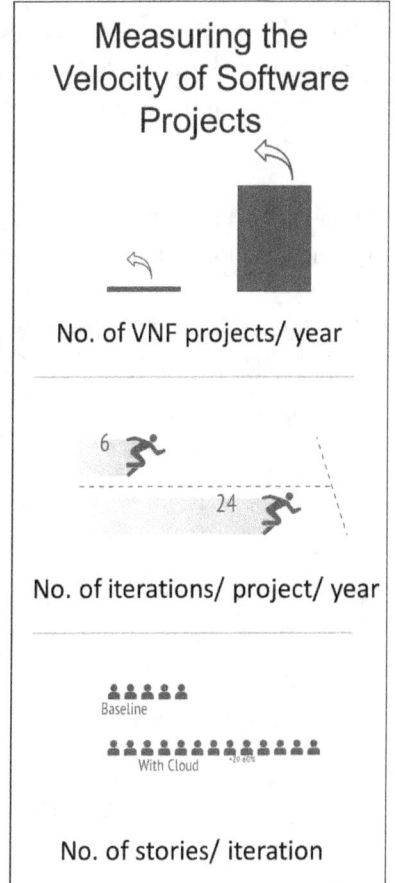

Some developers refactor cloud hosted VNFs to make them more cloud friendly, or "cloud optimized". A non-disruptive way to approach this effort is to take easily separable aspects of the monolithic application and convert them into services accessible via REST APIs. The VNF state may then be moved to a dedicated service, so the rest of the app becomes stateless. Making these changes allows for greater velocity in software development and the ability to perform cloud-centric operations such as scale-out, scale-in and self-healing.

While converting an existing VNF to be fully cloud native may be overly burdensome, all new VNFs should be written exclusively as cloud native if possible. (We have already covered cloud native application patterns in Chapter 2.) By using a cloud native architecture, developers and users can get much higher velocity in innovation and a high degree of flexibility in VNF orchestration and lifecycle management. In an enterprise end-user study conducted by Mirantis and Intel, the move to cloud native programming showed an average increase of iterations/year from 6 to 24 (4x increase) and a typical increase in the number of user stories/iteration of 20-60%. Enterprise cloud native apps are not the same as cloud native VNFs, but these benefits should generally apply to NFV as well.

Ultimately, there is no right or wrong architecture choice for existing VNFs (new VNFs should be designed as cloud native). The chart below shows VNF app architecture trade-offs.

Trade-offs between VNF Architectures

VNF Onboarding

The next major topic to consider when integrating VNFs into OPNFV scenarios is VNF onboarding. A VNF by itself is not very useful; the MANO layer needs associated metadata and descriptors to manage these VNFs. The VNF Package, which includes the VNF Descriptor (VNFD), describes what the VNF requires, how to configure the VNF, and how to manage its lifecycle. Along with this information, the VNF onboarding process may be viewed in four steps.

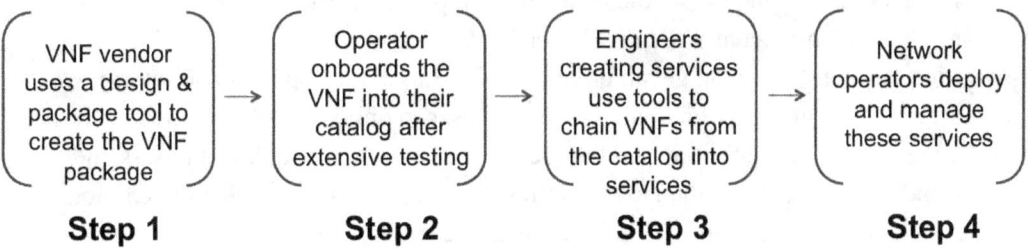

VNF Onboarding Steps

A detailed discussion of these steps is out of scope for this book and instead we will focus on the VNF package.

For successful VNF onboarding, the following types of attributes need to be specified in the VNF package. This list is by no means comprehensive; it is meant to be a sample. The package may include:

- Basic information such as:
    - Pricing
    - SLA
    - Licensing model
    - Provider
    - Version
- VNF packaging (tar or CSAR etc.)
- VNF configuration
- NFVI requirements such as:
    - vCPU
    - Memory
    - Storage
    - Data plane acceleration
    - CPU architecture
    - Affinity/anti-affinity
- VNF lifecycle management:
    - Start/stop
    - Scaling

- Healing
- Update
- Upgrade
- Termination

Currently, the industry lacks standards in the areas of VNF packaging and descriptors. Each MANO vendor or MANO project and each NFV vendor has its own format. By the time, you add VIM-specific considerations, you get an unmanageably large development and interop matrix. It could easily take months of manual work for a user to onboard VNFs to their specific MANO+VIM choice because the formats have to be adapted and then tested. Both users and VNF providers find this process less than ideal. Both sides are always wondering which models to support, and what components to proactively test against.

The VNF manager (VNFM) might further complicate the situation. For simple VNFs, a generic VNFM might be adequate. For more complex VNFs such as VoLTE (voice over LTE), a custom (read: proprietary) VNFM might be needed, and would be provided by the VNF vendor. Needless to say, the already complex interop matrix becomes even more complex in this case.

In addition to manual work and wasted time, there are other issues exposed by the lack of standards. For example, there is no way for a VNF to be sure it will be provided resources that match its requirements. There may also be gaps in security, isolation, scaling, self-healing and other lifecycle management phases.

OPNFV recognizes the importance of standardizing the VNF onboarding process. The MANO working group, along with the Models project (see Chapter 5) is working on standardizing VNF onboarding for OPNFV. The projects address multiple issues including VNF package development, VNF package import, VNF validation/testing (basic and in-service), VNF import into a catalog, service blueprint creation, and VNFD models. The three main modeling languages being considered are: UML, TOSCA-NFV simple profile, and YANG:

- **UML:** The Unified Modeling Language (UML) is standardized by the Object Management Group (OMG) and can be used for a variety of use cases. ETSI is using UML for standardizing their VNFD specification. At a high level, UML could be considered an application-centric language.
- **TOSCA-NFV simple profile:** TOSCA is a cloud-centric modeling language. A TOSCA blueprint describes a graph of node templates, along with their connectivity.

Next, workflows specify how a series of actions occur, which can get complex when considering various dependencies. Finally, TOSCA also allows for policies that trigger workflows based on events. The TOSCA-NFV simple profile specification covers an NFV-specific data model using the TOSCA language.
- **YANG:** YANG is a modeling language standardized by IETF. Unlike TOSCA or UML, YANG is a network-centric modeling language. YANG models the state data and configurations of network elements. YANG describes a tree of nodes and relationships between them.

OPNFV is considering all three approaches, and in some cases hybrid approaches with multiple modeling languages, to solve the VNF onboarding problem. Given the importance of this issue, there is also considerable collaboration with outside organizations and projects such as ETSI, TMForum, OASIS, ON.Lab, and so on.

Clearwater vIMS on OPNFV

Clearwater is a virtual IP multimedia system (vIMS) software VNF project, open-sourced by Metaswitch. It is a complex cloud native application with a number of interconnected virtual instances.

Clearwater vIMS

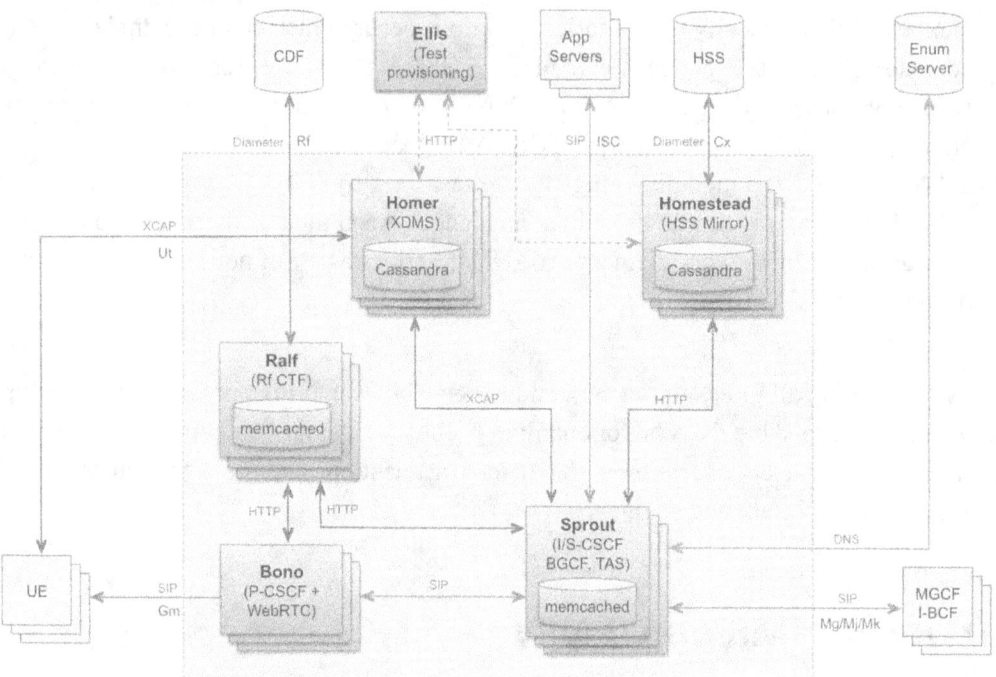

For OPNFV testing, TOSCA is used as the VNFD modeling language. The TOSCA blueprint first describes each of the nodes and their connectivity. A snippet of this code is shown below:

VNF Descriptor for Homestead HSS Mirror

```
homestead_host:
  type: clearwater.nodes.MonitoredServer
  capabilities:
    scalable:
      properties:
        min_instances: 1
  relationships:
    - target: base_security_group
      type: cloudify.openstack.server_connected_to_security_group
    - target: homestead_security_group
      type: cloudify.openstack.server_connected_to_security_group
homestead:
  type: clearwater.nodes.homestead
  properties:
    private_domain: clearwater.local
    release: { get_input: release }
  relationships:
    - type: cloudify.relationships.contained_in
      target: homestead_host
    - type: app_connected_to_bind
      target: bind
```

Next, the TOSCA blueprint describes a number of workflows. The workflows cover full lifecycle management of Clearwater. Finally, the blueprint describes policies that trigger workflows based on events.

Clearwater TOSCA Workflows

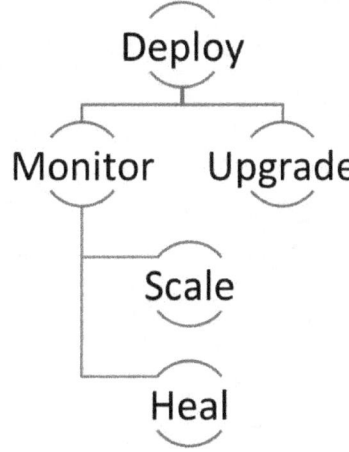

The TOSCA code fragment below shows a scale up policy based on a threshold, that then triggers a workflow to scale up Sprout SIP router instances from the initial 1 to a maximum of 5.

TOSCA Scaleup Policy and Workflow for Sprout SIP

```
policies:
  up_scale_policy:
    type: cloudify.policies.types.threshold
    properties:
      service: cpu.total.user
      threshold: 25
      stability_time: 60
    triggers:
      scale_trigger:
        type: cloudify.policies.triggers.execute_workflow
        parameters:
          workflow: scale
          workflow_parameters:
            scalable_entity_name: sprout
```

```
        delta: 1
        scale_compute: true
        max_instances: 5
```

Once the blueprint is complete, an orchestrator needs to interpret and act upon the TOSCA blueprint. For purposes of testing Clearwater, OPNFV uses Cloudify, a MANO product from Gigaspaces available in both commercial and open source flavors. Cloudify orchestrates each of the workflows described in the above blueprint. Specifically, the workflow to deploy the VNF looks like this:

Initial Deployment of the Clearwater VNF

[Diagram: Steps 4-9: Install & Configure VNF software. VNFs shown as DNS, Homestead, Homer, Sprout, Bono, Ellis over NFVI (KVM, Ceph, OVS, hardware etc.). MANO: Cloudify connected to VNFD TOSCA blueprint. MANO performs: 1. Create security groups, 2. Create VMs, 3. Create floating IP. VIM: OpenStack and SDN Controller shown.]

Running this entire series of steps in an automated fashion in Functest requires the following:

Step 1: Deploy VIM, SDN controller, NFVI

Step 2: Deploy the MANO software (could be Heat, Open-O or Cloudify, which is the current choice). For testing purposes, it is possible to use the full MANO stack (NFVO + VNFM) or just the VNFM.

Step 3: Test the VNF. For project Clearwater, Functest runs more than 100 default signaling tests covering most vIMS test cases (calls, registration, redirection, busy, and so on).

We have talked about a specific VNF, but this approach is pragmatic enough to be applied to other VNFs – open source or proprietary. Using OPNFV as a standard way to onboard VNFs brings great value to the industry because of the complexity of the VNF onboarding landscape. No one vendor or user has the resources or time to perform testing against a full interop matrix. But as a community, this is eminently possible.

At this point, it is worth taking a bit of a detour to illustrate the power of open source. The initial project Clearwater testing work was done by an intern at Orange. The work became quite popular, and has been adopted by numerous vendors, influenced the OPNFV MANO working group, and even convinced some operators to use OPNFV as a VNF onboarding vehicle.

In summary, we saw how VNFs can target different application architectures, what is involved in onboarding VNFs, and a concrete example of how the Clearwater vIMS VNF has been onboarded by OPNFV for testing purposes. In the next chapter, we will discuss how you can benefit from and get involved with the OPNFV project.

10
UTILIZE OPNFV TO DRIVE BUSINESS

The OPNFV project provides real business benefits to NFV end users, technology providers and individuals alike.

OPNFV and End-users

OPNFV provides both tangible and intangible benefits to end users. Tangible benefits include those that directly impact business metrics, while intangibles include benefits that speed up the overall NFV transformation journey but are harder to measure.

Tangible Benefits

Telecom operators and enterprises gain measurable business benefits by getting involved with the OPNFV project:

- **Faster rollouts of new services:** Networking deployments are often marred by endless delays from technology evaluations, proof-of-concept projects, vendor selections, field trials and production deployment. By getting involved with OPNFV, end users can complete many of these above steps in the community in collaboration with other end users. What remains is simply choosing a vendor whose products are based on an OPNFV stack, since these products benefit from all the community testing and interoperability work. This results in a direct time-to-market reduction for the end user.

- **VNF onboarding platform:** There is no easy way to compare benchmarks across VNFs since the underlying NFVI, SDN controller and VIM platform is not consistent across vendors. By using an OPNFV scenario, end users can standardize on a consistent environment against which different VNFs can be tested, characterized and benchmarked. Moreover, users can utilize Yardstick and other OPNFV performance testing projects to validate platform performance in-house. This results in both a time-to-market reduction and an improved technology stack resulting a higher project return-on-investment (ROI) for the end user.
- **Best-in-class components:** One of the major benefits of open source software, as we discussed in Chapter 3, is to be able to pick and choose best-in-class components aligned with specific use cases. OPNFV deliberately provides multiple choices at different layers of the NFV stack enabling this selection; this in turn results in a greater project ROI by increasing the value of the stack and freeing up valuable developer resources for other activities.
- **No vendor lock-in:** Another major benefit of open source software is the ability to eliminate vendor lock-in. Assuming the market is large enough, there are bound to be multiple vendors; and this drives down cost and makes it possible for users to switch vendors. We are by no means minimizing the effort of switching vendors, but it is significantly smaller than what would be required when switching across proprietary technologies. This results in a clear total-cost-of-ownership (TCO) reduction.
- **Product that solves real problems:** Rather than communicating with a proprietary vendor's engineering team through an army of gatekeepers such as account executives, systems engineers, product managers, systems architects, field CTOs, and so on, in the OPNFV community, users can work side-by-side with engineers from a vendor to ensure that features they care about are included in the product. Or users can contribute features themselves. This results in an improved project ROI for end users.

Intangible Benefits

It would be myopic to consider OPNFV as just technology. By getting involved with the OPNFV community, telecom operators can gain insights into all aspects of NFV transformation to speed up their journey.

Specifically, let's revisit the four pillars of NFV transformation discussed in Chapter 2, and evaluate how OPNFV helps with each one.

Impact to Organizational Structure

The 45+ project teams in OPNFV are small groups that consist of a PTL (project technical lead), committers and contributors. These teams prioritize their own backlog, develop requirements, contribute upstream code if required, develop their own tests, and ensure the readiness of their project for a given release. The technical steering community (TSC), infrastructure, and test projects perform coordination and common activities across OPNFV projects.

Users can learn how to organize their teams for successful NFV transformation by getting involved with OPNFV, and model their organizational structure after that of OPNFV. Let's revisit the organizational structure readiness questionnaire from Chapter 2 and see how OPNFV helps:

Question	Readiness: 1-5 (1 = Not ready, 5 = Completely ready)	Why?
Has there been any example where cross-functional groups formed teams for new technology?	4	Most teams have broad representation from both users and technology providers.
Can you imagine completely autonomous teams per NFV service or component (such as NFVI, VIM, MANO) that own dev, test, release **and** production?	5	Teams own all aspects. (Production phase is not applicable to OPNFV.)
Can you imagine a decentralized decision-making approach where each team, within reason, could make their own technical decisions?	5	This principle is fully incorporated into OPNFV. Each team, within reason, paves their own path.
Can you imagine the communication between NFV services teams and the platform team to be completely automated?	4	The interface between OSS/BSS and MANO, and MANO and NFVI/VIM is completely specified through APIs. No human interaction is required.
Are there mechanisms to discuss requirements that affect multiple	5	With the TSC, infra and test teams, there are plenty of

teams, and to coordinate these requirements?		mechanisms to coordinate between projects.

Impact to Process

While OPNFV does not manage a production NFV cloud, the development, test and release processes are closer to those of Web 2.0 companies than a traditional telecom operator. Decision-making and coordination between projects in OPNFV is governed by light-weight processes. For example, a proposal to create a new project is as simple as creating a Wiki page and presenting the project to the technical community after two weeks of initial review. If there are no objections, the proposal moves to the TSC for approval.

At a more concrete level, the CI/CD process employed by OPNFV provides a great opportunity for traditional telecoms to be exposed to DevOps methodologies that are sometimes thought to be inapplicable to the telecom community.

Users can learn how to retool their processes for successful NFV transformation by getting involved with OPNFV. Let's revisit the process readiness questionnaire from Chapter 2 and see how OPNFV helps:

Question	Readiness: 1-5 (1 = Not ready, 5 = Completely ready)	Why?
Can you imagine an intern pushing a change to a production network on their first day at work? (Etsy requires their interns to do so)	4	OPNFV does not have a CD pipeline yet, but an intern could submit a patch for CI on their first day.
Could you imagine an engineer experimenting with a new feature on the production network on 1% of your user-base without seeking any permission? (Twitter allows this.)	N/A	Not applicable.

Do you use Agile methodology? Is the methodology truly Agile or is it *waterscrumfall* (that is, a hybrid of waterfall and Agile)?	5	OPNFV methodology is fully agile, driven by epics and user stories in JIRA.
Can your finance team deal with hardware purchases not allocated to a specific cost center, but for an aggregate NFV cloud?	N/A	Not applicable.
Can your compliance, asset inventory, and security teams deal with a dynamic virtual infrastructure?	N/A	Not applicable.

Impact to Technology

The primary objective of the OPNFV project is to give users open source technology they can use and tailor for their purposes.

Over and above the benefits from the *Tangible Benefits* section above, by getting involved, users can also gain valuable experience with technologies required for a successful NFV transformation.

Let's revisit the technology readiness questionnaire from Chapter 2 and see how OPNFV helps:

Question	Readiness: 1-5 (1 = Not ready, 5 = Completely ready)	Why?
Are you ready for cloud native applications and microservices?	5	OPNFV uses OpenStack as a VIM, which is perfectly suited for cloud native applications.
Would your organization be receptive to the principles of Chaos Engineering?	4	While there isn't a formal Chaos Engineering project in OPNFV, the technical community culture is

		geared toward these same principles.
Is your organization ready to set up DevOps pipelines and automated testing?	5	DevOps CI/CD methodologies are the backbone of OPNFV.
Can your operations team adapt to model driven architectures?	5	OPNFV integrates open source MANO projects that are all model driven.
Can you imagine a monitoring framework where you cannot access single VNFs or instances, but rather the management is done on the aggregate cloud?	5	See Chapter 5, multiple projects assist with this aspect of managing the cloud in aggregate.

Impact to Skill Set Acquisition

By far the least expensive and most scalable method for a user to acquire the relevant skills for NFV transformation is to get involved with OPNFV. Just by joining the various modes of community communications, using the tools, and engaging with fellow community members, your team will gain both specific skills pertaining to the VIM, SDN controller, MANO, NFVI, service assurance, performance benchmarking, and so on, and also general cloud native skills such as model driven architectures, microservices and DevOps. There simply aren't enough formal classes in colleges or corporate training institutes to fill the skill set acquisition gap using traditional techniques.

Let's revisit the skill set readiness questionnaire from Chapter 2 and see how OPNFV helps:

Question	Readiness: 1-5 (1 = Not ready, 5 = Completely ready)	Why?
Are employees permitted any time and/or access to lab resources to learn new things on their own?	N/A	OPNFV Pharos Labs allow members and others to integrate with the CI/CD infrastructure and run test deploys of VNFs.

Can you imagine a company-wide Hack Week (week long Hackathon)?	5	OPNFV routinely conducts week long Hackfests and Plugfests
Are there formal training programs? How easy is it for technologists at your company to attend relevant trade shows or summits?	N/A	Not applicable.
Are there internal workshops or knowledge transfer programs? Will teams share learnings with each other?	4	OPNFV conducts summits and events to transfer knowledge, and there are many learning resources (mostly community-driven).
Are there post-mortems on projects? Can you imagine an environment where upper management will not assign blame during these post-mortems?	4	Each project conducts its own evaluations. The TSC does not assign any blame for missed deadlines, for example.

Sample RFP OPNFV Questions

It is not too early to include OPNFV-related questions in your request for proposals (RFPs). Here are some sample questions; we are sure your procurement team can come up with more.

Question
Does your product use any upstream components from the OPNFV project? If so, which ones?
Does your product incorporate any of the OPNFV upstream contribution projects? If yes, which ones? If not, why not?
Does your product incorporate any of the OPNFV testing and integration projects? If yes, which ones? If not, why not?
Does your product leverage any particular OPNFV scenario?
Do you use one of the installers used in OPNFV? If yes, which one? If not, please describe your installer.
Does your product test against OPNFV components? Explain.

Do you test against all OPNFV test suites? If yes, please provide results. If not, why not?
What value have you created over and above an OPNFV integrated stack in terms of functionality or services?
Does your internal process use continuous integration and testing?
Has your stack passed the OPNFV compliance verification program (not available yet, but coming soon)?

OPNFV and Technology Providers

In addition to being useful to users, OPNFV is also beneficial to technology providers such as:

- Silicon vendors (CPU, NIC etc.)
- Server manufacturers
- Switch & bare metal switch vendors
- Storage vendors
- VNF, MANO, SDN controller, VIM vendors
- Operating System and hypervisor vendors

Vendors creating products based on one of OPNFV's integrated stacks can shrink their schedules and reduce their effort by piggybacking on the efforts of the OPNFV community. Alternatively, vendors may choose to integrate their proprietary products with OPNFV. In these situations, OPNFV does not include proprietary products in their routine testing, but there is nothing to prevent technology providers from conducting their own testing with OPNFV by setting up a private Pharos-like lab and a CI pipeline to run specific test cases against a set of scenarios. Both the tests and scenarios can be enhanced to accommodate proprietary products. Another forum to test products is OPNFV plugfests, where all technologies and commercial products are welcome.

In both cases above, the ROI for the vendor's project improves significantly.

OPNFV and Individuals

Just about every activity in OPNFV is open to individuals as well. Getting involved with OPNFV is a mechanism to build the skills required for your next job at absolutely no cost. Skills around NFV are difficult to find, and the demand will only grow with market growth and the emergence of new use cases such as 5G. Even if you are not a developer, you can help with documentation, prioritizing pain points, refining use cases, and much more.

Getting Involved

There are many ways for users, technology providers and individuals to get involved with OPNFV. You don't have to be an OPNFV member to contribute, but many have decided to join and support open source NFV. The specific step to get involved requires a free Linux Foundation account. It's that simple!

Here are some specific ways to get involved, and the associated benefits:

Activity	Suitable For			Benefit
	End-user	Technology Provider	Individual	
Join existing project	x	x	x	Learn and make the stack more suitable for your use case
Create new project	x	x	x	Make the stack more suitable for your use case
Attend events/ Hackfests	x	x	x	Learn, collaborate, connect
Present at events	x	x	x	Share learning with others and recruit contributors
Continuously consume OPNFV using a CI pipeline	x	x	x	Learn and provide feedback to improve the stack
Contribute a Pharos Lab	x	x		Advance OPNFV testing capabilities
Join End-User Advisory Group	x			Provide feedback and requirements to ensure NFV pain points get solved

Present customer success stories	x			Help bring other users on board
Join plugfests		x		Ensure interoperability of your open or proprietary product(s) with OPNFV
Set up a "private" Pharos-like lab	x	x		Ensure interoperability of your proprietary code or product with OPNFV
Publish benchmarks & test results		x		Market your technology within the OPNFV context

In summary, OPNFV brings solid benefits to end users, technology providers and individuals.

The next generation network is being built today, and it's being built with open source tools. OPNFV is accelerating the testing, integration, and collaboration necessary to leverage open source NFV end-to-end.

The foundation of the next-generation network is forming; and OPNFV is accelerating the testing, integration, and collaboration necessary to leverage open source NFV end-to-end. It is easy to get involved, so why wait? Get involved today!

11
ADDITIONAL RESOURCES

General

OPNFV website	opnfv.org
OPNFV wiki	wiki.opnfv.org
OPNFV questions	ask.opnfv.org
OPNFV software & documentation	opnfv.org/software/downloads

Chapter 1 Resources

Telstra PEN	telstraglobal.com/products/connectivity/pen
The virtualization revolution: NFV unleashed video	youtu.be/-c5P2CvBTRg
Intel vCPE TCO calculator	builders.intel.com/docs/vE-CPE-for-communication-service-providers.pdf
ETSI NFV ISG	etsi.org/technologies-clusters/technologies/nfv

ETSI NFV architectural framework	etsi.org/deliver/etsi_gs/NFV/001_099/002/01.02.01_60/gs_NFV002v010201p.pdf
ETSI NFV terminology	etsi.org/deliver/etsi_gs/NFV/001_099/003/01.02.01_60/gs_NFV003v010201p.pdf
ETSI NFV POC framework	etsi.org/deliver/etsi_gs/NFV-PER/001_099/002/01.01.02_60/gs_NFV-PER002v010102p.pdf
ETSI NFV Security Problem Statement	etsi.org/deliver/etsi_gs/NFV-SEC/001_099/001/01.01.01_60/gs_NFV-SEC001v010101p.pdf
Heavy Reading Study on CSPs and OpenStack	openstack.org/assets/pdf-downloads/OpenStack-survey-results-public-presentation.pdf
IETS Network Service Header (Service Chaining)	datatracker.ietf.org/doc/draft-ietf-sfc-nsh/

Chapter 2 Resources

12 factor applications	12factor.net/config
About waterscrumfall	waterscrumfall.org/
Chaos engineering	principlesofchaos.org/
SDx Central on carriers embracing trial & error approach	sdxcentral.com/articles/news/sdn-world-congress-carriers-strive-agile-nfv-becomes-real/2016/10/

Chapter 3 Resources

OPNFV dashboard	opnfv.biterg.io

OPNFV end-user advisory group	opnfv.org/end-users/end-user-advisory-group
Communications in the new era of open by Heavy Reading	www.fujitsu.com/us/Images/Communications-in-the-New-Era-of-Open.pdf
CORD	opencord.org
VMware vSphere ESXi	vmware.com/products/vsphere.html

Chapter 4 Resources

ONAP	onap.org
OPEN-O	open-o.org
OPNFV Opera	wiki.opnfv.org/display/PROJ/OPNFV-OPEN-O
Juju	jujucharms.com
OpenBaton	openbaton.github.io
OPNFV Orchestra	wiki.opnfv.org/display/PROJ/Orchestra
OpenStack Tacker	docs.openstack.org/developer/tacker
Open Source MANO	osm.etsi.org
OpenStack	openstack.org
OpenStack telecom & NFV	openstack.org/telecoms-and-nfv
OpenStack project navigator	openstack.org/software/project-navigator
Kubernetes	kubernetes.io

OpenStack Neutron	wiki.openstack.org/wiki/Neutron
OpenDaylight	opendaylight.org
ONOS	onosproject.org
OpenContrail	opencontrail.org
OVN	benpfaff.org/~blp/dist-docs/ovn-architecture.7.html
KVM	linux-kvm.org
LXD	ubuntu.com/cloud/lxd
libvirt	libvirt.org
Ceph	ceph.com
OVS	openvswitch.org
FD.io	fd.io
FD.io performance study	lightreading.com/nfv/nfv-tests-and-trials/validating-ciscos-nfv-infrastructure-pt-1/d/d-id/718684?page_number=8
DPDK	dpdk.org
DPDK performance study	software.intel.com/en-us/articles/using-open-vswitch-with-dpdk-for-inter-vm-nfv-applications
ODP	opendataplane.org
Open Compute Project	opencompute.org

Chapter 5 Resources

Nokia: OpenStack with Real-Time Applications	openstack.org/videos/barcelona-2016/nokia-openstack-with-real-time-applications
Discussion about OVS DPDK showing 10x performance gain	wiki.opnfv.org/display/ovsnfv/Project+Proposal
Wikipedia fog computing page	en.wikipedia.org/wiki/Fog_computing
OPNFV CII Best Practices Badge	bestpractices.coreinfrastructure.org/projects/164
OPNFV Doctor	wiki.opnfv.org/display/doctor/Doctor+Home
Fault Management with OpenStack Congress and Vitrage, Based on OPNFV Doctor Framework video	openstack.org/videos/barcelona-2016/fault-management-with-openstack-congress-and-vitrage-based-on-opnfv-doctor-framework
OpenStack Keynote: OpenStack and OPNFV – Keeping Your Mobile Phone Calls Connected	youtube.com/watch?v=Dvh8q5m9Ahk&t=9s

Chapter 6 Resources

OPNFV JIRA	jira.opnfv.org
OPNFV project meetings	wiki.opnfv.org/display/meetings
OPNFV git	git.opnfv.org
OPNFV gerrit	gerrit.opnfv.org
OPNFV gitbhub mirror	github.com/opnfv

OPNFV Jenkins	build.opnfv.org
OPNFV artifact repository	artifacts.opnfv.org
OPNFV Docker hub repository	hub.docker.com/search/?q=opnfv
OPNFV Pharos	opnfv.org/community/projects/pharos
OPNFV Pharos dashboard	labs.opnfv.org

Chapter 7 Resources

OPNFV Danube scenarios	wiki.opnfv.org/display/SWREL/Danube+Scenario+Status
OPNFV Fuel	wiki.opnfv.org/display/fuel/Fuel+Opnfv
OPNFV Apex	wiki.opnfv.org/display/apex/Apex
OPNFV JOID	wiki.opnfv.org/display/joid/JOID+Home
OPNFV Compass	wiki.opnfv.org/display/compass4nfv/Compass4nfv

Chapter 8 Resources

OPNFV testing overview	wiki.opnfv.org/display/testing/TestPerf
Functest	wiki.opnfv.org/display/functest/Opnfv+Functional+Testing
Yardstick	wiki.opnfv.org/display/yardstick/Yardstick
VSPERF	wiki.opnfv.org/display/vsperf/VSperf+Home

Cperf	wiki.opnfv.org/display/meetings/CPerf
Storperf	wiki.opnfv.org/display/storperf/Storperf
Qtip	wiki.opnfv.org/display/qtip
Bottlenecks	wiki.opnfv.org/display/bottlenecks/Bottlenecks
Dovetail	wiki.opnfv.org/display/dovetail/Dovetail+Home
Danube release testing dashboard	testresults.opnfv.org/reporting/danube.html
Swagger API dashboard (creds opnfv/api@opnfv)	testresults.opnfv.org/test/swagger/spec.html
Grafana dashboard	testresults.opnfv.org/grafana
Functest dashboard	testresults.opnfv.org/reporting/functest-danube.html
OPNFV Dec'16 Plugfest report	opnfv.org/wp-content/uploads/2017/02/OPNFV_Plugfest_II_Report_FINAL.pdf

Chapter 9 Resources

Object Management Group	omg.org
ETSI VNFD specification	etsi.org/deliver/etsi_gs/NFV-IFA/001_099/011/02.01.01_60/gs_NFV-IFA011v020101p.pdf
TOSCA NFV simple profile	docs.oasis-open.org/tosca/tosca-nfv/v1.0/tosca-nfv-v1.0.html

YANG	tools.ietf.org/html/rfc6020
Project Clearwater	projectclearwater.org
Clearwater TOSCA Blueprint	github.com/Orange-OpenSource/opnfv-cloudify-clearwater
Cloudify	getcloudify.org

Chapter 10 Resources

OPNFV get involved	opnfv.org/community/get-involved
OPNFV tutorial videos (how to get involved)	youtube.com/playlist?list=PLiM029E-zvrs15SWfjfcJ19FWCsTT49f0

www.ingramcontent.com/pod-product-compliance
Lightning Source LLC
Chambersburg PA
CBHW081150180526
45170CB00006B/2011